Discovering Industrial Ecology

An Executive Briefing and Sourcebook

Ernest A. Lowe
John L. Warren
Stephen R. Moran

Library of Congress Cataloging-in-Publication Data

Lowe, Ernest A.
Discovering industrial ecology : an executive briefing and sourcebook / by Ernest A. Lowe and Stephen R. Moran, John L. Warren.
p. cm.
Includes bibliographical references.
ISBN 1-57477-034-9 (alk. paper)
1. Industrial ecology. 2. Industrial management—Environmental aspects. 3. Industries—Environmental aspects. I. Moran, Stephen R. II. Warren. John L. III. Title.
TS161.L68 1997
658.4´08—DC21 97-20763
CIP

Printed in the United States of America

Battelle Press
505 King Avenue
Columbus, Ohio 43201-2693
614–424–6393; 1–800–451–3543
Fax: 614–424–3819
Homepage: http://www.battelle.org/bookstore
e-mail: press@battelle.org

♲ Printed on recycled paper.

Acknowledgments

Based on a report prepared for:
Future Studies Unit
Office of Policy, Planning and Evaluation
U.S. Environmental Protection Agency
Washington, DC 20460
and the U.S. Department of Energy
under Contract DE-AC06-76RLO 1830

This sourcebook was based on a collaborative effort initiated by Dave Rejeski of EPA's Future Unit. The concept was initially developed by Ernest Lowe, who was the primary author. He was assisted by an Indigo Development team that included: Stephen Moran, Laurence Evans, Douglas B. Holmes, Grace Lowe, and Scott Butner. Other reviewers and contributors included: Brad Allenby, Faye Duchin, Suren Erkman, Gil Friend, Nicholas Gertler, Nancy Osborn, Deanna Richards, Kathleen Victory and Rose M. Watt—Pacific Northwest National Laboratory—design and page layout.

Professor Christine Rosen, Jumbi Edulberhram, Matt Owens, and Tracie Reynolds provided their UC Berkeley Business School class project, Industrial Ecology for Firms and Managers.

Professor Catherine Koshland, Dara O'Rourke, and Lloyd Connelly provided a valuable early draft of their critical review of industrial ecology (UC Berkeley Energy and Resources Group).

Dedication

This work is dedicated to the loving memory of the ecologist on our team who always kept us honest—
Dr. Lawrence K. Evans.

Thank you, Coyote!

How to Contact Us

Ernest A. Lowe, Indigo Development
A Division of RPP International
6423 Oakwood Dr.
Oakland, CA 94611-1362
510-339-1090 Fax: 510-339-9361
E-mail: elowe@indigodev.com
Web: www.indigodev.com

John L. Warren, Battelle
3230 Q Street
Richland, WA 99352
509-372-4759 Fax: 509-372-4370
E-mail: jl_warren@pnl.gov
Web: www.estd.battelle.org

Stephen R. Moran, Indigo Development
A Division of RPP International
3612 114A Street
Edmonton, Alberta, Canada
T6J 1N3
E-mail: smoran@aol.com

Using This Book

The executive briefing on industrial ecology (IE) conveys a quick overview of the field of industrial ecology. What is it? What activities does it encompass? What are the potential benefits of applying it? What business opportunities does it open? What strategies can we follow to evolve IE?

The second section is a sourcebook for readers wishing to learn more about the field and how they might apply it in their businesses and organizations. Here you will find further discussion of business opportunities, information on methods and tools typical of IE, cases illustrating IE principles, and scenarios of potential applications.

The cases illustrating IE are from many different fields of innovation—energy efficiency, pollution prevention, total quality environmental management, or often from applied common sense. Very few developed from anyone saying, "Let's do industrial ecology." But they suggest the results one might gain by consciously applying this new way of thinking.

The future scenarios are *hypothetical cases* illustrating a more ambitious level of application of industrial ecology. We include both public and private sector scenarios, ranging from the story of a local community using IE to that of a developing country.

There are new resources for industrial ecology available every day. For an update on available resources, consult either of these web sites: www.indigodev.com or www.estd.battelle.org.

Contents

Each section of the sourcebook also has references, organizations, and electronic addresses specific to its topic.

The Source of Value

. . . ON THE ROLE OF BIODIVERSITY AND NATURAL CAPITAL IN HUMAN ACTIVITIES

Natural systems are the source of value, the ultimate foundation of all economic values. This simple but often ignored fact is made clear by the notion of natural capital. Ecological economists such as Herman Daly suggest that earning profits on financial capital is only possible because of the flow of natural resources and services. Economic and industrial leaders must devote the same level of effort to keeping this natural capital intact that they devote to more traditional capital (the human means of production) (Daly and Cobb 1989, p. 71).

Natural capital includes regulatory functions such as maintaining balance in the chemical composition of the atmosphere and oceans, formation of topsoil and maintenance of its fertility, and maintenance of biological and genetic diversity. It also includes processing human wastes by soil, water, and air systems.

Natural capital production functions include providing raw materials for food, fiber, construction, energy, and medicine. The full list of benefits is much longer (de Groot 1994, pp. 151-168; Daily 1997, pp. 3-4).

"In this 'ecological perspective,' a separate and necessary (but not sufficient) condition for sustainability is the maintenance of an adequate 'environmental resource endowment'— the environmental assets necessary to provide needed and wanted environmental services.

"The most critical environmental services include the basic conditions of life-support on the earth, namely climate stabilization (temperature, rainfall, etc.), food supply (the 'food chain'), and biological waste disposal and materials recycling.

"It is noteworthy that climatic stability, moderate temperatures, the stratospheric ozone layer, the carbon-oxygen cycle, the nitrogen cycle, the hydrological cycle, mature 'old growth' forests (temperate or tropical), soil fertility and bio-diversity are not technologically replaceable (or repairable) to any meaningful degree, at least for the planet as a whole."

(Ayres 1991)

Ecologists emphasize that humanity enjoys the goods and services of this natural capital thanks to the integrity of ecosystems and larger ecological cycles. They also point out that we are just beginning to understand how sensitive is the balance in these systems and how vulnerable they are to human activities. Furthermore, ecologists agree that the diversity of living species is fundamental to healthy ecological processes.

Thus, biodiversity is a central requirement for keeping the Earth's natural capital intact. Natural goods and services come not from machines but from living systems whose viability depends upon maintaining a high variety of species and redundancy of function. We humans need to recognize that our own long-term survival is intertwined with the biological diversity of these systems.

RESOURCES

Ayres, Robert U. 1991. "Industrial Metabolism: Closing the Materials Cycle." Paper prepared for SEI Conference, Principles of Clean Production, Stockholm.

Daily, Gretchen C. 1997. *Nature's Services: Societal Dependence on Natural Ecosystems*. Island Press, Washington, D.C.

Daly, Herman and Cobb, John. 1989. *For the Common Good: Redirecting the Economy Toward Community, the Environment, and a Sustainable Future*. Beacon Press.

de Groot, Rudolf S. 1994. "Environmental Functions and the Economic Value of Natural Ecosystems." In Jansson, AnnMari, et al. (eds.). *Investing in Natural Capital: The Ecological Economics Approach to Sustainability*. Island Press, Washington, DC.

Discovering Industrial Ecology

An Executive Briefing and Sourcebook

1 An Executive Briefing on Industrial Ecology

THE CHALLENGE

The level of resource consumption and environmental degradation involved in our current economic/industrial system is one of the major challenges facing the world in the 21st century. This challenge is compounded by the rapid growth of many emerging economies. The following quote from a seminal paper on industrial ecology, published in 1989 by Robert Frosch and Nicholas Gallopoulos, summarizes the challenge:

> By the year 2030, 10 billion people are likely to live on this planet: ideally, all would enjoy standards of living equivalent to those of industrial democracies such as the U.S. or Japan. If they consume critical natural resources at current U.S. rates . . . and if new resources or substitutes are not discovered, such an ideal would last a decade or less. On the waste side of the ledger, at current U.S. rates, 10 billion people would generate 400 billion tons of solid waste every year—enough to bury greater Los Angeles 100 meters deep (Frosch and Gallopoulos 1989, p. 144).

The standard of living of industrialized societies is based on an economic/industrial system that consumes huge

"In all essential respects, down to the tiniest molecule, we humans have the same structure as endangered birds of prey, seals, and otters. In a biological sense we are neither the masters of nature nor its stewards, but a piece of nature ourselves, just like seals and otters.

And if these species have become threatened with extinction in the space of a few decades because of our environmental pollution, we too are threatened. We have enough knowledge to say that the only way of reversing this process is to avoid introducing substances into nature that it cannot process and to learn to live cyclically just like cells in nature."

(Robért 1993)

amounts of materials and energy on a flow-through basis. Robert Ayres provides a quantitative description of this system in the U.S. :

> Ten tons of active mass raw materials (not including construction materials) per person [are] extracted from U.S. territory by the economy. Roughly 75 percent is mineral and nonrenewable while 25 percent is, in principle, from renewable sources. Six percent of the total is embodied in durable products. The other 94 percent is converted into waste residuals as fast as it is extracted (Ayres 1989).

The U.S. economy has very low energy system efficiencies in terms of the services actually delivered to society. Massive waste defines potential business opportunities and directions for policy development. The total input of energy to transportation in the U.S. equals the total energy demand of the Japanese economy, 20 quadrillion BTUs (quads). *18.4 quads of U.S. transportation energy* is wasted in heat dissipated from engines, vehicles idling in gridlock, and at other points along the flow. This enormous level of wasted energy suggests both fundamental redesign of products and of transportation systems. U.S. power plants also match the size of total Japanese energy usage in the waste heat they send up the stacks: 20 quads.

A system based on throughput of resources is viable if resources and sinks for waste materials are unlimited. When the population grows to a level where its rate of consumption of resources exceeds the availability of resources, the organism involved is starved. When the generation of waste materials exceeds the capacity of the sinks, the environment is poisoned and the organism perishes. Human economies are beginning to run out of resources and sinks for our wastes.

The challenge to already industrialized and emerging economies is to create sustainable industrial systems. The source of value for economic systems must be transformed from maximizing throughput and consumption of resources to optimizing quality of life within the constraints of natural systems. This requires integration of human economic activity and material management with biological, chemical, and physical global systems. Industrial ecology offers a framework for achieving this transition.

THE VISION

Meeting these challenges requires a vision to guide the evolution of the industrial system. Proposed elements of this vision include:

Industry operates within the limits of global, regional, and local carrying capacity, maintaining a cautious margin for error.

Industry reflects ecological and biological principles in the design and operation of its activities, from the shop floor to the executive suite.

Materials are cycled through the economy to an optimal degree, approaching a closed-loop system.

Major energy sources are renewable and solar based.

Use of renewable materials is in balance with their production.

Nonrenewable materials are conserved and valued.

Diversity of life is maintained as a foundation for the viability of the whole system of life.

Efficiency and productivity are in dynamic balance with resiliency, assuring continued capacity to adapt to change.

Societies make the transition to this state while maintaining the economic viability of systems for extraction, production, distribution, transportation, and services.

The transition supports development of more viable communities, with an improved quality of life around the planet.

"Model the systemic design of industry on the systemic design of the natural system . . . industrial ecology involves designing industrial infrastructures as if they were a series of interlocking man-made ecosystems interfacing with the natural global ecosystem . . . .

"The aim of industrial ecology is to interpret and adapt an understanding of the natural system and apply it to the design of the man-made system, in order to achieve a pattern of industrialization that is not only more efficient, but which is intrinsically adjusted to the tolerances and characteristics of the natural system."

(Tibbs 1992)

DEFINING INDUSTRIAL ECOLOGY

Industrial ecology is both a context for action and a field of research. We focus on its implications for action and the need for connections among action, policy, and research.

Industrial ecology is a dynamic systems-based framework that enables management of human activity on a sustainable basis by:

- Minimizing energy and materials usage
- Ensuring acceptable quality of life for people
- Minimizing the ecological impact of human activity to levels natural systems can sustain
- Maintaining the economic viability of systems for industry, trade, and commerce.

"Industrial ecology is a systems oriented approach to integration of human economic activity and material management with fundamental biological, chemical, and physical global systems. Industrial ecology is a means by which our species can deliberately and rationally approach and maintain a desirable global carrying capacity."

(Allenby 1992)
(Braden Allenby is AT&T Vice-President for Technology and the Environment.)

The industrial ecology approach involves (1) applying systems science to industrial systems, (2) defining the system boundary to incorporate the natural world, and (3) seeking to optimize that system. In this context, "industrial systems" apply not just to private sector manufacturing and service, but also to government operations, including provision of infrastructure.

Industrial ecology (IE) is a context for designing and operating industrial systems as living systems interdependent with natural systems. It provides a framework for balancing environmental and economic performance with an emerging understanding of local and global ecological constraints. It supports coordination of design over the life cycle of products and processes. IE enables creation of short-term innovations in a long-term context.

While much of the initial work in IE has focused on manufacturing, a full definition of "industrial systems" includes service, agricultural, manufacturing, and public operations, as well as infrastructure such as landfills, water and sewage systems, and transportation systems.

FIVE THINGS TO REMEMBER ABOUT INDUSTRIAL ECOLOGY

1. Industrial systems are living systems that operate in living systems.
2. Industrial ecology opens new opportunities for business.
3. Industrial ecology opens new opportunities for government.
4. Strategies for creating and implementing IE are emerging.
5. Methods and tools are available for applying industrial ecology.

1. Industrial Systems Are Living Systems That Operate Within Living Systems

Industrial ecology's definitive contribution is moving the vision of industry from a linear, mechanistic model to a closed-loop system, more akin to natural ecosystems. Achieving this major objective can be supported by perceiving a manufacturing plant, a municipal sewage

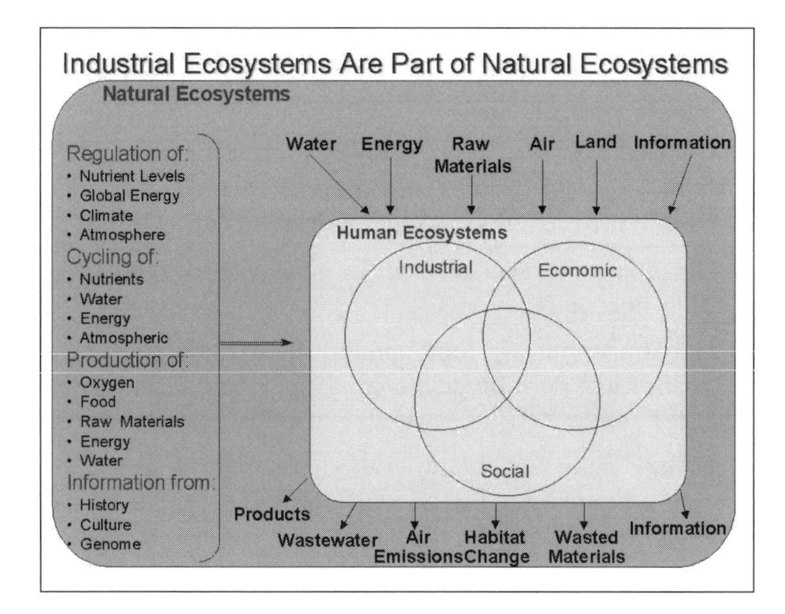

treatment plant, or a hotel as a living system, a part of larger natural systems.

In our present, linear system we emphasize moving materials and energy through the economic system as the primary source of value. Producing and consuming are the dominant economic activities.

Compare this to ecosystems where production and decomposition are well-balanced, with nutrients recycling continuously to support the next cycles of production. In an ecosystem, species diversity assures redundancy in functions essential to the whole, cooperation is in balance with competition, and evolution of the system occurs through self-organization, not top-down control. These characteristics contribute to resiliency, enabling the system to adapt to both major and minor changes in conditions over time.

The industrial system will need this resilience of living systems to make a successful transition to a sustainable economy.

Systems science developed out of the interdisciplinary study of living systems. So there is a large body of experience, methods, and theory to support viewing industry as a living system, embedded in ecological systems. Industrial systems are living systems, for example:

Applying IE at Three Levels:

The following illustrates how IE could be applied at different levels of design and through different scales of time.

In transportation, an industrial ecologist would support short-term enhancements in automobile design through such tools as design for environment. The basic question would be: how can we optimize trade-offs to reduce energy use and pollution in the production process as well as during use of the product?

At another level, an industrial ecologist might ask, how can we transform small vehicle design to capture levels of efficiency and freedom from pollution not possible within the internal combustion model?

At a still broader level, an industrial ecologist (possibly in public policy or a competing business) would ask, how can we design integrated transportation systems to move people with the highest resource efficiency and lowest possible pollution? How can telecommunications, urban planning, and work design reduce the number of trips and distance traveled?

(See the end of Chapter 3 for a detailed discussion of these three levels of design.)

"Find pollution or waste and you've found something you paid for but can't sell. You've found inefficiency . . . fundamentally all pollution is lost profit . . . by striving to eliminate it, we together can grow a more efficient, competitive economy."

—Peter Coors, CEO
Coors Brewing Co.

Novo Nordisk has captured over 50 percent of the billion dollar world market for industrial enzymes. This competitive advantage came from early recognition that biological substitutes for synthetic chemicals would be in high demand (see Chapter 5 for details).

- ◆ Industrial systems (at any level) are dependent upon and impact local and regional ecosystems and the planetary biosphere.
- ◆ Material inputs to our economy ultimately come from natural ecosystems. (Over geological time, minerals continually cycle through living systems.)
- ◆ The major portion of energy now comes from *fossil* fuels, once living plants.
- ◆ The atmosphere (a fundamental industrial resource) is the product of biological cycles operating on a global scale.
- ◆ Design, management, operation, regulation, and consumption are all in the hands of *homo sapiens*, a diverse collection of living creatures.

2. Industrial Ecology Opens New Opportunities for Business

The transition to an industrial ecology perspective provides opportunities for existing and new businesses. Companies' profitability can benefit from increased resource efficiencies, new products, and new services. IE is likely to transform our understanding of competitive advantage, efficiency, and even corporate mission.

Industry may find the impact of this ecological transformation of business and public operations equal to that of the continuing information revolution. For instance, some companies are already shifting their mission from maximizing sale of products to delivering services to the market. *See sidebar on Agfa-Gevaert and Chapter 4.*

Firms will find new applications for old technologies and markets for new ones. For instance, companies manufacturing equipment for materials and chemical processing will find new customers in intermediary firms reprocessing wastes into feedstocks.

IE's systems viewpoint may enable entrepreneurs to enter markets now dominated by traditional technical solutions. In energy, for instance, photovoltaic cells and films, when fully integrated into building components (including roof tiles, facades, and glazing), are almost competitive with fossil fuel energy sources. In a developing country, an innovative company might gain contracts through design of a decentralized renewable energy system integrating use of wind, photovoltaics, and biomass sources with smaller, fossil fuel power plants.

In Chapter 5 we discuss opportunities suggested by IE, including:

- ◆ Initiating cost savings and new revenues in existing operations
- ◆ Finding new markets for existing goods and services
- ◆ Marketing new technologies, materials, and processes
- ◆ Supporting the organizational change, technical, and information needs of an IE-based economy
- ◆ Integrating technologies and methods into innovative new systems
- ◆ Consulting, training, and information services supporting firms and public organizations in making the transition.

The Swiss photocopier company, Agfa-Gevaert, demonstrates a systems shift in business mission that reduces demand on material and energy resources. AGt leases copiers in Switzerland with a long-term flexible agreement (selling system utilization) which covers all consumables in a price per copy. The company assumes responsibility for product quality and utility over its lifetime. Therefore designers have a strong incentive to use long-life components, standardize components and systems, lower costs of supplies, and aim for ease of repair and reconditioning. "Agfa-Gevaert's interest in product durability is evident: the longer its products can be rented out and the cheaper their operation is, the higher its profit."

(Borlin 1990, see Chapter 4)

3. Industrial Ecology Opens New Opportunities for Government

Industrial ecology principles, methods, and tools can support government in discovering paths to sustainability. IE can assist agencies at all levels of government in defining policy and regulations, organizing agencies to work more effectively in partnership with the private sector, guiding and financing research and development, and facilitating sustainable economic development.

Industrial ecology can support the work of public policy makers and regulators who are seeking more effective, systemic approaches to environmental protection, technology policy, and economic development. For example:

- ◆ IE's global view offers an appropriate context for prioritizing risks and identifying points of high leverage for change.
- ◆ IE methods such as industrial metabolism and dynamic input-output modeling provide means for assessing alternative policy options.
- ◆ Organizational structures for implementing policy can also benefit from IE's strategy of learning from the dynamics and principles of ecosystems, particularly their processes of regulation and self-regulation.

"We have thousands and thousands of environmental regulations on the books. Too often, our environmental activities have been compartmentalized, law by law, pollutant by pollutant. And too often, instead of preventing pollution, we have simply shuffled and shifted pollution from one place to another—from land to air, from air to water, from water to land."

—Carol Browner, Administrator, U.S. EPA

- IE can assist policy making in energy, transportation, agriculture, and service industries as well as in manufacturing.
- Public sector facilities and systems managers can improve efficiency and increase the return on public investment by applying industrial ecology in their operations. These range from military bases to local energy, water, sewage, and solid waste systems.
- IE can help economic development policy makers and managers form more effective strategies for sustainable development.
- Local economic development can benefit from the many business opportunities opened by IE.

Regulators face twin challenges: creating more systemic, market-based, voluntary approaches to enfranchise high-performance corporations and, simultaneously, remaining able to enforce compliance by more recalcitrant firms.

Over the past several years, both government and business have acknowledged the limitations of the material-specific, medium-specific, command and control approach to environmental protection. Industrial ecology is a valuable foundation for efforts to reform policy and regulations.

Industrial ecology provides a means of balancing environmental and economic objectives. Industrial ecologists dismiss the myth that there is inherent conflict between the two. IE's perspective that industry is a part of nature opens new, creative policy and program options for government.

Industrial ecology can support the charting of a path to sustainability for large, economically crucial industries that are high in demand on non-renewable resources and generation of pollution. It is a framework in which public/private partnerships can better coordinate strategic planning across short- to long-term futures.

See Chapter 6 for a discussion of the government's role in IE.

4. Strategies for Creating and Implementing IE Are Emerging

Strategies for the development and application of industrial ecology must, of course, reflect ecological principles.

Ecosystems are self-organizing. Control in them is decentralized and redundant.

Industrial ecologists propose that similar strategies will be fundamental to evolving this field. No one individual or organization can possess the knowledge to provide centralized, top-down planning and control. The strategy for the transition from a linear, mechanical model of industry to a closed-loop, living system economy must emerge through a self-organizing process.

Four key thrusts for this strategy are emerging:

1. Implementation of many decentralized demonstration projects in diverse industrial and public organizations
2. Application of IE principles in more central institutions, the points of social and economic control, such as the World Bank, regulatory and development agencies, and standards groups
3. Continuing research on the interactions between industry, the environment, and society; action research on demonstration projects; and development of new IE methods and tools
4. Rapid dissemination of information about IE principles and methods and about the successes and failures of the demonstration projects.

The systems view of IE requires a strategy that coordinates action within individual firms, across networks of firms, and between industry, the market, and the government. Several basic principles will support strategic directions for developing IE:

- Coordinate a decentralized, self-organizing process of innovation with strong information links across systems and levels of system.
- Engage the points of social and economic control in the process of developing IE.
- Design policy through public/private, highest common denominator dialogue, framed by IE understandings and project experiences.
- Use the leverage of industries at risk from environmental degradation (e.g., insurance, finance, and tourism) to support change in the sectors most responsible.

Self-Organizing Systems

Humans have a strong capacity for managing their own work or exploration. The most dramatic examples are the responses to earthquakes or floods where people start rescue work before the professionals arrive. Jury duty is a more ordinary case of self-organization. With no organizational guideline beyond "elect a foreperson," a jury organizes itself to make often difficult decisions.

Systems scientists see self-organizing processes as fundamental to the successful operation of any endeavor. No top-down controller can know enough about the many details within operational units to set the way they organize to get their tasks done.

The development path for industrial ecology will be a complex system, requiring self-organizing strategies as one foundation.

"What the earth needs most is a variety of useful models—model homes, buildings, companies, communities, and countries, all demonstrating how to make the transition from linear to cyclical processes. Positive examples are an extremely powerful force for change, and it takes only a small proportion of a population—perhaps as little as 15 percent—to stimulate dramatic improvements."

—Karl-Henrik Robért is founder of The Natural Step initiative which started in Sweden.

- Maintain a strong linkage between people and institutions applying IE and scientists researching IE.
- Develop industry's leadership role in changing consumer behavior and expectations.
- Build upon present transforming trends in industry.

The World Wide Web has become a powerful resource in support of the development and dissemination of IE, linking research centers, policy development, and action projects.

See Chapter 8 for discussion of these points.

5. Methods and Tools Are Available for Applying Industrial Ecology

IE provides a context for creating new methods and tools and for applying old ones in new ways. Its view of industry as a living system operating in nature brings a larger perspective for expanding the effectiveness of existing approaches to management.

IE enables evolution of the present frameworks for designing and managing industrial systems that generally assume them to be linear, mechanical systems, largely uncoupled from nature. A powerful array of tools to enable change to a closed-loop, cyclical system is emerging from the thinking embedded in industrial ecology and related fields.

Industrial ecologists have used a basic industrial tool called design for x (DFX) as the foundation for a systemic approach to environmental issues in product or process design. This is called design for environment (DFE). AT&T has led an electronics industry task force in developing and applying DFE (see Chapter 3).

Environmental accounting practices such as full cost accounting and resource accounting are changing the way companies and nations look at the relationship between the economy and the environment. These tools replace economic models and financial accounting systems that define environmental impacts as externalities and undervalue or ignore natural capital (resources and capacity to absorb industrial wastes).

Design tools such as life-cycle assessment (LCA) and design for environment (DFE) are providing breakthrough improvement in environmental performance and creating a sustainable competitive advantage for the companies employing them. These tools replace a system in which design decisions are made solely on the basis of financial and technical factors, with environmental issues treated as externalities to be controlled at the end of the pipe.

Logistics engineering offers a major opportunity to adapt tools designed for managing military systems to the task of developing and operating an industrial ecosystem of companies exchanging by-product materials.

Full application of industrial ecology suggests many changes in the structures of organizations and information systems. So methods for organizational design and change management will be an important support to implementation processes. Advanced environmental information systems will play a major role within and between organizations.

To achieve their objectives, industrial ecologists will also need to develop new methods and tools; for instance, ways to coordinate the planning of local, incremental changes with long-term organizational and societal objectives.

Development of IE methods will also require building bridges with related fields such as ecological economics and sustainable business management.

From an IE perspective, there are a number of requirements for the design of methods and tools that enable managers and designers to

- Coordinate planning and action across time and space (local to global, short- to long-term)
- Link incremental changes into broader transformative change
- Coordinate redesign of industrial systems with emerging understanding of the environmental challenges we face
- Balance economic and environmental considerations (human needs and ecological needs and constraints)
- Balance efficiency and resilience in system design.

Faye Duchin, an economist working in the IE context, has evolved static industrial input-output models into a dynamic 'what-if' tool. Dynamic input-output modeling enables corporate or public policy makers to consider the implications of alternative policies and strategies.

Through this lens they can see the interactions between economic, technical, and environmental variables. For instance, this process could test a policy for fostering renewable energy as the foundation for a developing country's energy needs (see Chapter 3).

A CONVERSATION AFTER THE BRIEFING[1]

A manufacturing executive (ME), a manufacturing plant manager (MPM), a service company executive (SCE), a city public works director (CPWD), and an environmental agency executive (EAE) gather after the briefing to discuss and challenge what they've heard with an industrial ecologist (IE).

[1]The systems scientist Stafford Beer pioneered the use of a fictional conversation to challenge his own ideas in *The Heart of Enterprise*, John Wiley & Sons, 1979, New York, NY. This book is a comprehensive description of the viable system model for organizational management.

MPM: This industrial ecology really isn't anything new. It's just a fancy buzzword for what we're already doing!

IE: You're mostly right at your level. Industrial belt tightening has prompted lots of plant managers like you to look for cost savings through waste minimization, process changes, internal and external recycling, and other pollution prevention tactics.

What we believe industrial ecology offers is a more strategic and integrated approach, helping you move beyond incremental, within-the-plant improvements. Take the pulp and paper industry for instance. Advanced forestry methods may improve performance at the input end. Pollution prevention and energy efficiency get results in the mills. Business and consumer recycling help close the loop. Industrial ecology lets you work to optimize the whole system and find the best trade-offs between those areas.

If you're modeling a hotel, you'll see that it eats, breathes, drinks, and eliminates. Since this living system also has consciousness (in your managers, employees, and customers) it is capable of learning and changing to improve its metabolism.

SCE: I'm still struggling with this idea of a hotel or a fast-food chain being a living system embedded in natural systems. Intuitively it seems to make sense. But how do we put that into operation?

IE: One starting point would be doing a qualitative model of the "metabolism" of one of your operating units. Look at the inputs, the processes through which you deliver service, and the outputs. Include the elements that don't involve a financial transaction as well as those that do. Then look at the natural systems that those inputs, outputs, and processes impact. Keep the model simple at first. Build detail as you ask new questions and find new answers.

ME: But the regs just aren't written that way, and they are the main drivers of our environmental behavior.

EAE: I guess that's why I'm here at this briefing. We're struggling too, trying to take thirty years of fragmented policies and regulations and reframe them into this systems way of thinking. In our best initiatives we're trying to work with industries and give you guys the space you need to innovate.

We want you to be able to come up with better answers than any we could ever prescribe. You know your operations and your technologies, so our best role should be helping to set the targets for change and keeping information flowing. Is that industrial ecology or is it just common sense?

CPWD: A lot of what you're saying about IE *is* just common sense. That seems kind of appealing to me. I look at my landfill that's almost full up and count the costs of that! I see a big public investment. Spending a little money to get the whole city thinking "waste not, want not" could save us millions. And think of the jobs we could create turning wastes into feedstocks and products!

IE: I'm glad that you see the common sense side of this field. I think IE will help us discover many common sense solutions and put them together into the whole ball of wax. I believe that the systems view is itself common sense.

MPM: But you're talking about more than common sense. There are new tools to invent. New ways of thinking to learn. And it sounds as though we're going to be putting a lot of time and money into making something real out of a bunch of high-flown ideas.

IE: You're right again, MPM. That's why I begin by saying industrial ecology is a framework, a way of seeing and thinking. Thinking this way puts a spin on how we use established methods like TQM or concurrent engineering. Your designers already understand DFX, so it's not a big jump to design for environment. That limits your investment in change. IE is able to piggyback on a lot that you're already doing.

ME: But realizing that vision you hit us with is going to take more than tuning up existing methods and tools. You used the word "transformation," but I think it looks more like a revolution. I just don't see that IE gives us the means of achieving that level of change.

IE: That's why we need to develop new means, with IE as a *framework* for creating them. The toolkit definitely needs new methods and tools. This innovation has to come out of close alliances between industry, research, and public policy.

ME: That's where research consortia come in. I can team with other companies in my industry to do this R&D, pulling in the universities, the government labs, and the regulators. Working in this way cuts the investment my company has to make.

CPWD: I can see some ways that industrial ecology helps out people in public operations, but I've not seen much about us or service industries in the articles I've read. Are industrial ecologists thinking only about manufacturing industries?

IE: You've pointed out a very important gap. The basic concepts of IE can benefit any industrial operation, whether it be managing water and sewage flows, moving goods and people around the landscape, or producing those goods.

For instance, you and your city planners could develop a regional transportation plan based on IE objectives and principles. You would explicitly account for the impacts on ecosystems and regional natural systems while insuring effective and efficient movement for economic and personal travel.

ME: Are you sure IE isn't going to be used by the regulators to create a new club to hold over us? Our state got us involved in a voluntary program, we cut emissions below what the regs called for, and then the Feds came in with new mandatory performance standards. That took all flexibility out of our system.

EAE: Let me take the heat on that one. I know your case, and I know that was a real bad error. As little as I understand IE, I do get that it just won't work if we try to use it as a club. After twenty years as a regulator, I've discovered there is actually a systems science of regulation called cybernetics. It tells us to set policies and regs with respect for your knowledge of your business as well as our knowledge of natural systems. That has to be basic to how we apply industrial ecology.

SCE: What you say implies that *someone* still has to be setting objectives based on what we discover from the research side of IE. Is that going to be you regulators?

EAE: We're just part of the process. Setting objectives has to come out of real dialogue between government, industry, and research. As our industrial ecologist reminded us, ecosystems don't function through top-down control.

ME: The Natural Step in Sweden is one model for how that can happen through a process including strong industry and citizen input. That's an example of industrial ecology thinking that emerged spontaneously, without anyone ever using the phrase. Our company is looking at getting involved.

SCE: I've heard about another version of this process called the Green Plan. Norway, Canada, and the Netherlands have all gone through processes where industry works together with government and citizens to form a long-term strategic plan. Funny, they don't use the words "industrial ecology" either.

IE: Believe it or not, we industrial ecologists don't care what you call it! In a decade or so I hope all of the buzzwords are in the compost heap and we have a world where the companies, governments, and people are all doing the right thing just because it makes sense.

RESOURCES FOR THE EXECUTIVE BRIEFING

Allenby, Braden. 1992. "Industrial Ecology: The Materials Scientist in an Environmentally Constrained World." *The Materials Research Society Bulletin*. March, Pittsburgh, PA, pp. 46-51.

Ayres, Robert U. 1989. "Externalities: Economics and Thermodynamics." In Archibugi and Nijkamp (eds.). *Economy and Ecology*. Kulwer Academic Publishers, Netherlands.

Borlin, M. 1990. "Swiss Case-Studies of Product Durability Strategy," Product-Life Institute, Zurick.

Coors, Peter. Quote from "The Environment: The Road from Rio," a special advertising section appearing in *The Economist*.

Frosch, Robert A., and Gallopoulos, Nicholas E. 1989. "Strategies for Manufacturing." *Scientific American*, Special Edition, September, pp. 144-152.

Robért, Karl-Henrik. 1993. *The Natural Step—A Social Invention for the Environment*. The Swedish Institute, Stockholm.

Tibbs, Hardin. 1992. "Industrial Ecology, An Environmental Agenda for Industry." *Whole Earth Review #77*, Winter, pp. 4-19.

White, Robert M. 1994. Preface to Allenby, Braden R., and Richards, Deanna J. *The Greening of Industrial Ecosystems*. Washington, DC: National Academy Press.

A general list of resources, organizations, and references is at the end of the book. Each chapter includes resources specific to its content.

2 Industry as Living Systems Within Living Systems

INTRODUCTION

From an industrial ecology perspective, the activities of the human economy are living systems participating in the Earth's natural systems. This perception has led some industrial ecologists to emphasize the value of using an ecological metaphor to learn how human activities can be environmentally benign, if not positive.

At a simple level, they recommend learning from nature by applying ecological principles in designing human systems. A more systemic use of the metaphor would be modeling specific industrial systems upon specific ecosystem dynamics (i.e., model materials recycling on the cycle of decomposition in a forest).

Setting objectives for the industrial system is perhaps the most fundamental level of learning from nature. If global and local economies are to become sustainable, we will have to develop performance objectives based upon deep understanding of the interaction between industrial systems and larger natural systems. Unfortunately, the idea that nature sets real constraints upon human activity is still controversial.

What does it mean to view a factory or municipal utility as a living system? First, it has a metabolism, drawing materials and energy from other living (or once living) systems (forests, farms, fossil fuels, oceans, the atmosphere, etc.). It disseminates its products to humans and its unsold by-products to natural systems. It is designed, managed, operated, and regulated by homo sapiens. *The air that moves through its processes and employees is itself the result of the interactions of forests, oceans, and animal life.*

The ecological metaphor also underlies the concept of the industrial ecosystem (and the eco-industrial park). This is covered in Chapter 7.

INDUSTRIAL ECOLOGY AND SUSTAINABLE DEVELOPMENT

Simply stated, sustainable development seeks two basic objectives:

- Widely shared, high quality of life for humans, continuing through the generations
- Healthy, diverse local ecosystems and restored balance in global systems.

The challenge to our industrialized economies and societies is to create a transition from the present, very unsustainable system without major disruption and breakdown.

What is IE's unique role in achieving this transition? It provides a view of the whole system. Human systems, including industry, are seen as natural systems interacting with local ecosystems, regional environments, and the ecosphere. This integrative perception is embodied in methods and tools for design and management of human systems. *See the next chapter.* IE is closely allied with ecologically based innovations in economics, business management, public policy, architecture, and planning.

The principles and methods of industrial ecology will enable leaders, managers, and designers in private and public industrial systems to design the transition to sustainability for their organizations. For each role, the core questions are: *How will we keep our present operations viable while charting a course to future operations that realizes the two objectives set above? How do we gain and sustain profitability (or balanced budgets for public organizations) without continuing growth of resource consumption and throughput?*

Industrial ecology can help such leaders and designers to ask the right questions and to find the short-term answers that lead to the necessary long-term outcomes.

For instance, an agricultural system dependent upon petrochemical products has a limited future. Long before oil supplies run out, industrialized farms are demonstrating systemic damage from use of pesticides, herbicides, chemical fertilizer, and heavy equipment. Without an IE foundation, some agricultural supply companies have started to develop genetically engineered crops as the silver bullets to increase production and to reduce pollution and energy demand. Creation of truly sustainable agriculture requires a systems view to understand the proper use of such technologies.

Supplier companies can develop successful transition strategies by perceiving their customers' fields as ecosystems and using IE methods such as design for environment, industrial metabolism, and input-output modeling. Genetically engineered plants are only one of many solutions that must be integrated into a whole system, including ecological soil management, integrated pest management (using natural predators for pests), more efficient irrigation systems, sensors, and other information system tools.

Individual firms need support from social institutions in creating and implementing their transition strategies. Broad objectives, benchmarks, and timelines for the transition will emerge from public/private dialogue. Industry will develop guidelines in concert with academic research, industry standards groups, policy-setting agencies, and citizen participation. The Natural Step and Green Plans demonstrate how this process has started. *See The Natural Step later in this chapter and more on Green Plans in Chapter 6.*

The foundation concepts of sustainability are not complex: all human systems are interdependent with ecological systems; the purpose of industrial systems is to support the quality of life of humans while restoring and maintaining high environmental quality. The challenge to consumers as well as business and government is moving to a system in which quality of life is not dependent upon continual increases in resources used and products sold, owned, and thrown away.

"But ask now the beasts, and they shall teach thee; and the fowls of the air, and they shall teach thee; Or speak to the earth and it shall teach thee; and the fishes of the sea shall declare unto thee."

(Job 12:7-8 King James Version)

LEARNING FROM NATURE

Industrial ecology suggests two core principles for design and management: view each level of industry as a living system participating in larger natural systems, and use the principles and dynamics of ecosystems to guide industrial design.

This call for a new ecological metaphor reflects awareness of the limitations of earlier metaphors for human systems: the clock, the machine, military command and control, and the computer. None of these analogues enables adequate design of the relationship between industrial activity and the environment.

Modeling industrial systems design upon ecosystem design can help humans achieve a healthy and sustainable relationship with nature. Nature has demonstrated effective "design" strategies through 3.5 billion years of evolution. Knowledge of the principles and dynamics of ecosystems and larger natural systems is a valuable support to the redesign of our industrialized culture.

Some Useful Principles of Living Systems

Ecosystems in nature demonstrate many strategies relevant to the design of industrial systems. A few notable examples include:

- The sole source of power for ecosystems is solar energy.
- In natural systems there is no such thing as "waste" in the sense of something that cannot eventually be absorbed constructively somewhere else in the system. Nutrients for one organism are derived from the death and decay of another.
- A major portion of energy flows in ecosystems is consumed in decomposition processes that make nutrients in wastes available.
- Concentrated toxic materials are generated and used locally.
- Efficiency and productivity are in dynamic balance with resiliency. This balance preserves the ability of ecosystems to adapt and evolve.
- In the face of change, ecosystems remain resilient through high biodiversity of species, organized in complex webs of relationships. The many relationships are maintained through self-organizing processes, not top-down control.
- In an ecosystem, each individual in a species acts independently, yet its activity patterns cooperatively mesh

with the patterns of other species. *Cooperation and competition are interlinked and held in balance.*

We will explore how several of these principles are relevant to industrial activities. Major currents in business practice reflect ones like the balance between cooperation and competition. All aspects of organizations can benefit from applying ecological principles, not just design and production systems.

APPLYING ECOLOGICAL PRINCIPLES TO INDUSTRIAL SYSTEMS

Principle 1: In natural systems there is no such thing as "waste" in the sense of something that cannot be absorbed constructively somewhere else in the system.

In nature, decomposers are among the most important of all organisms. Without fungi, bacteria, termites, nematodes, and other flora and fauna, the world would be overwhelmed by waste. Instead, decomposers make efficient use of natural "waste" by recycling matter into nutrients and other organic compounds that can be absorbed anew by the system (Lowe 1993).

With the right metrics in place, one could calculate the cost to industry of producing waste. Components include:

- ◆ Raw material
- ◆ Processing/transforming the material
- ◆ Allocated overhead
- ◆ Embedded energy
- ◆ Disposal costs
- ◆ Potential liabilities.

Making these costs visible through total cost accounting would enable companies to determine a more realistic bottom line.

Pollution prevention (in its waste minimization mode) and recycling have prompted many companies to search for markets for their unutilized products in the last decade.

The U.S. subsidiary of the German chemical and film manufacturer, BASF, has set up an Environmental Opportunities initiative in its corporate Ecology Department. The mission is to find uses and markets for

Chaparral Steel is a Midlothian, Texas, company seeking to define every output as a product, embodying the principle that waste is a sacrificed financial opportunity. CEO Gordon Forward's strategic goal is to generate zero waste.

Chaparral's Project STAR (Systems and Technology for Advanced Recycling) sets targets for reduced resource consumption, enhancing the value of by-products and reducing waste volumes. The company estimates that this program has gained over $6 million in new revenue and saved $2.9 million per year. Specific improvements include

- ◆ *Baghouse dust: Process changes reduced disposal costs and increased recovery of metals, saving $2.9 million*
- ◆ *Slag value: Process changes increased the value of the by-product to $6 million (as a cement component)*
- ◆ *Solvents: Replacement with nontoxic products and reduction in volume saved $400,000*
- ◆ *Using other wastes: Water and energy.*

Chaparral estimates that it is gaining $6 million in new revenue and saving $2.9 million per year through these innovations.

(Quinn 1995)

residual products not currently marketed. The initiative seeks to integrate this practice into day-to-day operations as a way of moving beyond simple compliance to waste disposal regulations. The company develops new revenues while saving on disposal costs (*Business and the Environment*, March 1994).

US Gypsum is making 100 percent recycled content sheet rock from synthetic gypsum generated as a by-product of utility flue gas desulfurization. It has been using recycled paper since 1930 and claims to be the largest recycler of waste paper in the United States (*Business and the Environment*, October 1992, p. 9).

These examples suggest that the ecological dynamic of linking organisms in food webs where one's wastes are nutrients for others makes good business sense. This move beyond individual plant boundaries raises new design questions: *Do we need to modify processes in company A to turn residuals into a product useful in company B? What regulatory and liability issues do such transfers raise? Are we perhaps locking in the use of a hazardous material rather than designing it out of the system?*

Some interpreters of industrial ecology identify the application of this one principle, no waste in ecosystems, as the total definition of the field. The field is far broader than this.

Principle 2: Concentrated toxins are not stored or transported in bulk at the system level, but are synthesized and used as needed only by species individuals.

Examples of this principle in nature are venomous snakes and poisonous plants, which are very discrete repositories of toxins. (A major exception in nature is where volcanic eruptions may distribute toxins across a wide area. Here, wetlands ecosystems may absorb and ultimately filter large masses of such materials.)

Quad/Graphics seeks competitive advantage by holding a broad, multifunctional vision of its mission which includes but goes beyond profitability

(Quad/Graphics, Enviro/Facts, Pewaukee WI 1993)

In industry, AT&T is piloting online production of gallium arsenide for computer chips from chemical precursors instead of shipping this extremely toxic chemical (Graedel and Allenby 1995).

Principle 3: Each member of an ecosystem performs multiple functions as it interrelates with other members.

An example in nature: During its lifetime a single deer perpetuates its own species by reproducing, encourages new grass and other plant species to grow by grazing and browsing and fertilizing the soil with its waste, and may provide food and energy for a wolf, coyote, or cougar. After death, its remains are converted by decomposers back into the organic compounds that fuel the earth's essential biogeochemical cycles upon which all species depend for life (Lowe 1993).

Quad/Graphics, the largest privately owned printing company in the U.S., has aggressively led the industry in pioneering green printing processes and materials. Rather than hold what it has learned as proprietary, the company has embarked on an extensive educational program to encourage other players in the industry, its customers, and end-consumers to follow its lead.

Principle 4: Carrying capacity limits the extent to which populations can grow.

Ecologists always look at the natural limits in any ecosystem they study. Quality of soil and underlying geology, climate patterns, genetic diversity, and other natural limits define the extent to which plant and animal populations can grow.

Unfortunately, this central ecological concept remains controversial when applied to human activity. Technological optimists prefer to think that technical innovation will enable a pattern of continuing industrial growth. Many leaders in developing countries resent concerns about their population growth expressed by industrialized countries. However, some industrial ecology pioneers like Braden Allenby, Robert Ayres, and Faye Duchin make the concept of carrying capacity central to their thinking.

At the level of a company seeking a location for a new plant, considering the carrying capacity of the local resource base could avoid major difficulties. The plant design team needs to work with awareness of the natural limits of the local ecosystems at possible sites. John Warren of Battelle gives a graphic example: "A North Carolina paper mill is located in an area that is a great place for a mill *except* that it has more air inversions than any other place in the U.S. The air is capped three-fourths of the summer. Review of the carrying capacity of the area would have shown it was not a good location for a mill" (Warren 1994).

"There are two kinds of natural selection, or two aspects of the struggle for existence: organism versus organism, which leads to competition, and organism versus environment, which leads to mutualism. To survive, an organism does not compete with its environment as it might with another organism, but it must adapt to or modify its community in a cooperative manner."

(Odum 1986)

PRINCIPLES OF INDUSTRIAL ECOLOGY

Hardin Tibbs, an industrial design engineer who has played an important role in disseminating IE concepts, suggests a number of general principles for the field (Tibbs 1992 as adapted by Indigo Development for Lowe 1995).

The word "industrial" is used here to also denote service and construction, not just manufacturing industries.

Connect individual firms into industrial ecosystems

- ◆ Close loops through reuse and recycling.
- ◆ Maximize efficiency of materials and energy use.
- ◆ Minimize waste generation.
- ◆ Define all wastes as potential products and seek markets for them.

Balance inputs and outputs to natural ecosystem capacities

- ◆ Reduce the environmental burden created by releases of energy and material into the natural environment.
- ◆ Design the industrial interface with the natural world in terms of the characteristics and sensitivity of the natural receiving environment.
- ◆ Avoid or minimize creating and transporting toxic and hazardous materials (when needed, synthesize locally).

Re-engineer industrial use of energy and materials

- ◆ Redesign processes to reduce energy usage.
- ◆ Substitute technologies and product design to reduce use of materials that disperse them beyond possibility of recapture.
- ◆ Do more with less (technically called dematerialization).

Align policy with a long-term perspective of industrial system evolution.

LEARNING FROM THE DYNAMICS OF ECOSYSTEMS: RECYCLING AND DECOMPOSER/DETRITIVORE ECOSYSTEMS

The analogic approach of industrial ecology may yield more powerful results when practiced in terms of ecosystem dynamics and interactions. Rather than simply drawing isolated principles from ecology, designers can model ecosystems to create more effective complex industrial systems. Ecosystems are tested viable systems, evolved over millennia. They have great resilience in the face of challenges. Insights into how they maintain viability could help create and improve industrial systems in both financial and environmental terms.

The process would involve a dialogue between industrial system designers (at the appropriate level) and an ecosystem researcher, comparing notes, laying out models side by side, and asking, "How does nature handle this issue we're stuck with in our Chicago operation?"

The industrial system is most unlike ecosystems in its profound imbalance between production/consumption and recycling of materials. Comparative studies of natural and industrial recycling systems could further the transition to closed-loop systems.

Designers have developed a remarkable variety of innovations in recycling industrial, commercial, and consumer waste over the last two decades. These range from development of new products and processes using recycled feedstocks to sophisticated resource recovery equipment for processing many components of a community's waste stream. However, much of this innovation is fragmented by the materials or disciplines involved. With most materials, market demand is not keeping up with supplies.

Overall, we still have a socioeconomic system in which over 90 percent of extracted material ends up as unused waste. Most energy is focused in production and consumption. In ecosystems, materials are reused and as much as 90 percent of energy goes directly into the system of decomposition to continually renew the nutrients needed for ongoing life. Nature's recycling of plant detritus and animal carcasses and feces closes the loop. Without it, forests and prairies would be buried in waste.

A more systemic framework for recycling could be developed for prioritizing research, selecting among recycling options, and guiding public policy. The aim would be to help restore human systems to a balance more closely resembling that found in ecosystems.

An industrial association, a major waste management company, or a policy institute could recruit an interdisciplinary team including ecologists expert on decomposition, waste management and ecological engineers, policy makers, and business managers. They would model

A startling array of organisms are responsible for decomposition processes in an ecosystem: decomposer animal forms range from protozoa, nematodes (eelworms), turbellarians, millipedes, centipedes, earthworms, and wood lice. Larger vertebrates like woodpeckers also play a key role. The soil microflora are dominated by bacteria and fungi, although algae also play an important role.

The tasks decomposer organisms achieve include fragmenting dead matter, chemically processing wastes to elemental form, generating nutrients through resynthesis of elements, and combining nutrients into humus for the next cycle of growth.

the total pattern of "recycling" in one or more specific ecosystems while building a model of recycling in industrial and consumer systems. Typical questions in the inquiry would be:

What do the ecosystem dynamics in natural decomposition suggest for integration of recycling technologies into a more unified system?

What are the principal strategies for breaking down and reusing materials in natural systems that could inspire new processes for recycling society's wastes?

How do the specific roles of organisms and interactions among them suggest new technical innovations and their interrelations? Are there promising specific biological processes not now utilized in recycling and treatment?

How are the processes of decomposition integrated with the productive processes in ecosystems? What implications does this have for industrial systems of production and consumption?

GUIDANCE FROM NATURE IN SETTING OBJECTIVES

Up to the present, objectives for environmental performance of public and private industrial operations have largely been set by law. The resulting objectives are fragmented and sometimes mutually contradictory. They do not stem from a systemic understanding of the environment the laws aim to protect. Regulated objectives are just beginning to take into account the global impacts of human activity apparent in the last 15 years.

Industrial ecologists seek to form a more immediate connection between objective setting and a systems awareness of natural needs and constraints. IE emphasizes an active role for industry in this process, working in collaboration with government, researchers, and citizens.

The *Fieldbook for Development of Eco-Industrial Parks* includes an IE-based framework for setting performance objectives for an industrial park as a whole and for each company in it. (This document is available from Indigo

Development as *Eco-Industrial Parks, A Guidebook for Local Development Teams*.) The broad categories are

1. Optimize resource utilization within industrial processes
   - ◆ Energy
   - ◆ Water
   - ◆ Materials
2. Minimize releases from industrial processes
   - ◆ Liquid waste
   - ◆ Atmospheric releases
   - ◆ Solid waste
3. Harmonize interactions of industrial processes and releases with natural systems
   - ◆ Habitat/wildlife
   - ◆ Neighbors
   - ◆ Physical setting.

The use of this framework involves attention to global as well as local conditions, preservation of biodiversity, and quality of life. Participants would develop economic objectives in parallel with their environmental objectives. Performance targets would be reset regularly in light of new information about the environment and technology.

The Natural Step

Karl-Henrik Robért, a Swedish physician, has created an effective public process to build a national agreement on objectives parallel to those identified by industrial ecologists. The Natural Step program forms consensus reports on the state of the environment and strategies for change.

These reports are developed by networks of experts, including business leaders, scientists, economists, activists, and politicians. Through booklets, audio cassettes, study circles, and other media the reports are circulated throughout Sweden. The process asks four fundamental questions regarding human action:

The process seeks common ground for people of widely different interests and beliefs. Stakeholders reach agreement, in part, because The Natural Step emphasizes beginning with more basic facts and principles rather than the specifics that often block understanding. With this agreement it is easier to approach specific issues in their context.

- ◆ Does the action reduce dependence upon finite mineral and fossil fuel resources and the deposition of their residues in natural systems?
- ◆ Does the action reduce the use of long-lived synthetic products or molecules?

- Does it preserve or increase natural diversity and the capacity of ecological cycles?
- Does it increase the efficiency of use of energy and other resources?

Expert networks, including representatives of the industries under consideration, have formed consensus documents in energy, metals, transportation, and agriculture, among other sectors. Businesses are creating action plans based upon objectives agreed to in the documents. Municipalities are also forming action plans and exchanging information on their experiences in implementing them.

The Natural Step (TNS) International has established offices in seven countries beyond Sweden. A network of TNS trainers is available to work with companies and communities in the United States. See resources for contact information.

In the next chapter we summarize a number of measures of sustainability suggested by the work of Robert Ayres in industrial metabolism. These focus on efficiency of resource use as an objective.

See Chapter 6 for information on Green Plans, another process for setting objectives and designing environmental strategies.

LIMITS TO THE ECOLOGICAL METAPHOR

Use of the ecosystem metaphor will limit itself. Designers only need ask, Are we still getting useful insights by using the metaphor in specific industrial settings and in broader business or societal issues? Or do we need to move to other systems views?

The ecosystem metaphor can only go so far. In ecosystems the equivalent of a "planning function" is distributed, self-organizing, unconscious, evolutionary, and long-term (except in catastrophe). There's no conscious controlling process. So industrial ecology cannot stop with modeling upon nature but must also apply systemic models reflecting the unique qualities of human consciousness.

General systems theory and cybernetics grew from the search for principles bridging natural, human, and technical systems. This work laid the foundation for organizational design frameworks like systems dynamics, the viable system model, sociotechnical systems, the learning organization, and self-managing work teams. All can support the organizational transformation to sustainability.

Ecological awareness joined with systemic self-awareness is required to participate in this task of coevolution. Learning from analogy with ecosystem design is one mode of "doing" industrial ecology, not the whole of it. In the next chapter we will explore related approaches in industrial ecology. Then later we will return to the metaphor with consideration of "industrial ecosystems" and attempts under way to "design" them as eco-industrial parks.

INDUSTRIAL ECOLOGY AND SYSTEMS

As a branch of systems science and systems thinking, industrial ecology improves the understanding and management of industrial activities (broadly defined) as systems nested within the natural systems of our planet, its regions, and local ecosystems.

What are systems? What is systems science? These words are overused and often abused. This is a brief introduction to what we mean by them and how they relate to IE.

- A system is a set of elements interrelating in a structured way.
- The elements are perceived as a whole with a purpose.
- The elements interact within *defined* boundaries.
- The behavior of the system results from the interaction of the elements and between the system and its environment. (System + environment = a larger system.)
- The properties of a system emerge from the interaction of its elements and are distinct from their properties as separate pieces.
- A system's behavior cannot be predicted by analysis of its individual elements.
- The definition of the elements and the setting of system boundaries are *subjective* actions. Conflicts often develop because different parties have defined very different systems without realizing it. So the assumptions of the definers or observers of any system must be made explicit.

Systems science ranges from highly theoretical work defining research methods to applied work in virtually all areas of life (often called "systems thinking"). Some modes of applying systems thinking include the learning

"Restructuring a system doesn't mean shoving people or things around, bulldozing, rebuilding, hiring, firing–that's not what changes system behavior. Almost always, the most effective restructuring means putting information into a place where it doesn't now reach, or changing goals, rewards, incentives, and disincentives so that the same people, in the same positions, make decisions a different way. Restructuring a system means changing what's in people's heads."

(Meadows 1992)

organization, systems dynamics, sociotechnical systems, and the viable system model. In this time of complex and rapid change, systems thinking has immediate, pragmatic value for companies and agencies of any size.

An industrial ecology example of the subjective nature of systems definition follows. Understanding that we *construct* a system from a particular point of view is crucial to working with systems thinking and IE. This concept often helps to resolve conflicts between differing points of view.

Systems Thinking About Steel

Different firms and agencies define quite different systems relating to this basic commodity and its environmental impacts or benefits.

Managers in a mini-mill company are interested in such elements as the reliability of supply and costs of recycled scrap, technological breakthroughs that increase strength and durability, and process changes that lower emissions. For them, the purpose of the system may be to build a competitive edge for recycled steel.

Managers in an auto company, on the other hand, may define a system focused on materials selection that enables them to weigh the relative environmental, production, and cost factors of steel, aluminum, and plastic. They may choose design for environment or life-cycle analysis tools to gain competitive advantage for their product designs.

One important overlap between the steel producer's and the steel user's systems is competitive advantage. The more the mini-mill reduces the environmental impacts of steel while keeping prices right, the more desirable it will be in the final product.

An environmental or economic development agency may define a system seeking to optimize the total recycling of metals, including steel in a region or country. Elements could include relative recycling rates for different sources, industry needs for technical transfer and information flows, government procurement policy, and new business niches.

Making these different systems definitions and values explicit would support strategic planning in public/private partnerships or supplier/customer relations.

RESOURCES ON ECOLOGY AND LIVING SYSTEMS

Organizations and Electronic Sources

The International Institute for Sustainable Development (IISD) is a nonprofit organization established and supported by the Governments of Canada and Manitoba. IISD's mandate is to promote sustainable development in decision making within government, business, and the daily lives of individuals in Canada and internationally. IISD Web site-http://iisd1.iisd.ca.

The International Institute for Applied Systems Analysis (IIASA) has been the site for IM work by Ayres and other researchers. IIASA's address is A-2361 Laxenburg, Austria. Telephone, (43 02236) 71521-0, Fax: 71313, Internet: postmaster@iiasa.ac.at. The World Wide Web address is http://www.iias.ac.at/welcome.html. Information on the Rhine Basin IM study is available online here.

The Natural Step U.S.: 4000 Bridgeway Sausalito CA, 415-332-9394; e-mail: tns@naturalstep.org. The Swedish office: Det Naturliga Steget, Slottsbacken 6, Stockholm S-111 30, SWEDEN; Tel: 011 46 8 545 12 500; Fax: 011 46 8 545 12 589; pdahlberg@detnaturligasteget.se. Canada: kelly_baxter@msn.com. Australia: natstep@peg.apc.org. Netherlands: codename@euronet.nl or Fax: 011 31 70 364 49 34. UK: forum@forumlon.demon.uk. France: Fax: 011 33 5 62 68 86 91.

Bibliographic Resources

Allenby, Braden R., and Cooper, William E. 1994. "Understanding Industrial Ecology from a Biological Systems Perspective." *Total Quality Environmental Management*. Spring, New York.

Ayres, Robert U. 1991. "Industrial Metabolism: Closing the Materials Cycle." Paper prepared for SEI Conference, Principles of Clean Production, Stockholm.

Beer, Stafford. 1966. *Decision and Control*, John Wiley & Son, Chichester, NY, Brisbane, Toronto, pp. 33-36, 111-119. Contains valuable insights into using parallel systems models, as with the example here of drawing upon the dynamics of decomposition in an ecosystem to design more effective recycling industries described above.

Côté, Raymond P., et al. 1994. *Designing and Operating Industrial Parks as Ecosystems*. School for Resource and Environmental Studies, Faculty of Management, Dalhousie University. Halifax, Nova Scotia B3J 1B9. For more detailed discussion of ecological concepts in industrial design see pp. 5-16.

Daly, Herman. 1991. *Steady-State Economics*, 2nd edition with new essays, Island Press.

Daly, Herman, and Cobb, John. 1989. *For the Common Good: Redirecting the Economy Toward Community, the Environment, and a Sustainable Future*. Beacon Press.

de Groot, Rudolf S. 1994. "Environmental Functions and the Economic Value of Natural Ecosystems." In Jansson, AnnMari, et al. (eds). *Investing in Natural Capital: The Ecological Economics Approach to Sustainability*. Island Press, Washington, DC, pp. 151-168.

Erkman, Suren. 1995. *Ecologie industrielle, métabolisme industriel, et société d'utilisation*. A major study supported by the Foundation for the Progress of Humanity, Paris. Dr. Erkman has gathered information on the evolution and application of industrial ecology in Europe, North America, and Asia. An English translation is being prepared.

Frosch, Robert A. 1992. "Industrial Ecology: A Philosophical Introduction." In National Academy of Sciences *Proceedings*, February.

Frosch, Robert A., and Gallopoulos, Nicholas E. 1989. "Strategies for Manufacturing. Scientific *American* (Special Edition, September), pp. 144-152.

Graedel, T.E., and Allenby, B.R. 1995. *Industrial Ecology*. Prentice Hall, Englewood Cliffs, NJ.

Hall, Charles S., Cleveland, Cutler J., and Kaufmann, Robert. 1986. *Energy and Resource Quality: The Ecology of the Economic Process*. John Wiley & Sons, New York. The expression "industrial ecosystems" and, more importantly, the full concept of industrial ecology, appears in the Preface, on page xiv. It is probably one of the very first occurrences of this expression in the published literature.

Hawken, Paul. 1994. *Ecology of Commerce: A Declaration of Sustainability*. Harper Business, New York, NY.

Holmberg, John. 1995. *Socio-Ecological Principles and Indicators for Sustainability*, Institute of Physical Resource Theory, Chalmers University of Technology and Göteborg University, S-412 96 Göteborg Sweden. This publication contains the theoretical foundation for the Natural Step process. E-mail: frt@fy.chalmers.se.

Kelly, Kevin. 1994. *Out of Control: The Rise of Neo-Biological Civilization*. Addison-Wesley, Reading, MA. Review the broad movement toward human and technical systems based on biological models, including industrial ecology, artificial life in computer science, complexity theory, and other emerging fields.

Lowe, Martha. 1993. *Natural Principles for Industrial Ecology*. Unpublished paper.

Meadows, Donella, Meadows, Dennis, and Randers, Jorgen. 1992. *Beyond the Limits to Growth: Confronting Global Collapse, Envisioning a Sustainable Future.* White River Junction, VT: Chelsea Green Publishers.

Miller, James Grier. 1978. *Living Systems*. McGraw-Hill Book, New York, NY.

O'Riordan, Timothy. 1994. "The Precaution Principle in Environmental Management." In Robert U. Ayres and Udo E. Simoni (eds.), *Industrial Metabolism. Restructuring for Sustainable Development*. United Nations University Press, Tokyo, New York, pp. 299-318.

Pizzocaro, Silvia. 1994. "Theoretical Approaches to Industrial Ecology: Status and Perspectives." Draft for discussion. Dipartimento PPPE, Politecnico di Milano, via Bonardi 3, 20133 Milano, Italia. Tele: 39 2 2399.5124, Fax: 39 2 2399.5150. An in depth exploration of the ecological metaphor in IE.

Quinn, Barbara. 1995. "Redefining Recycling: Everything is a Product." *Pollution Engineering*, October.

Rees, W., and Wackernagel, M. 1994. "Ecological Footprints and Appropriated Carrying Capacity: Measuring the Natural Capital Requirements of the Human Economy." In *Investing in Natural Capital: The Ecological Economics Approach to Sustainability*. A-M Jannson, M. Hammer, C. Folke, and R. Costanza (eds.), Island Press, Washington, DC.

Robért, Karl-Henrik. *The Natural Challenge* is scheduled for publication in the U.S. in 1996.

Socolow, R., Andrews, C., Berkhout, F., and Thomas, V. (eds.). 1994. *Industrial Ecology and Global Change*. Cambridge University Press, New York, NY.

Tibbs, Hardin. 1992. "Industrial Ecology, An Environmental Agenda for Industry." *Whole Earth Review* #77, Winter, pp. 4-19.

Todd, Nancy Jack, and Todd, John. 1994. *From Eco-Cities to Living Machines*. North Atlantic Books. A new edition of *Bioshelters, Ocean Arks, City Farming: Ecology as the Basis of Design*. 1984. Sierra Club Books, San Francisco. The Todd's work in biological design at New Alchemy Institute and Ocean Arks International has pioneered exploration of many themes in industrial ecology.

Warren, John. 1994. "Pollution Prevention Methods: Extending the Concept to Industrial Ecosystems." Proceedings, *Industrial Ecology Workshop, Making Business More Competitive*. Ontario Ministry of Environment and Energy, Toronto, Canada.

Watanabe, Chihiro. 1993. "Energy and Environmental Technologies in Sustainable Development: A View from Japan." *The Bridge*. Summer. National Academy of Engineering, Washington, DC, Summer. Sketches Japanese work in industrial ecology beginning in 1972.

"The idea of an industrial ecology is based upon a straightforward analogy with natural ecological systems. In nature an ecological system operates through a web of connections in which organisms live and consume each other and each other's waste. The system has evolved so that the characteristic of communities of living organisms seems to be that nothing that contains available energy or useful material will be lost. There will evolve some organism that manages to make its living by dealing with any waste product that provides available energy or usable material. Ecologists talk of a food web: an interconnection of uses of both organisms and their wastes. In the industrial context we may think of this as being use of products and waste products. The system structure of a natural ecology and the structure of an industrial system, or an economic system, are extremely similar."

(Frosch 1992)
(Robert Frosch, Kennedy School, Harvard, was Vice President in charge of General Motors Research Laboratories until 1993.)

For a General Introduction to Ecology

Ehrlich, Paul R. 1986. *The Machinery of Nature, The Living World Around Us—And How It Works*. A Touchstone Book, Simon & Schuster, Inc., New York.

Pavlik, Bruce, Muick, Sharon, et al. 1992. *Oaks of California*. Cachuma Press, Los Olivos, CA. An excellent introduction to ecology through study of a specific type of ecosystem.

Some works by ecologists useful as foundation for industrial ecology

Holling, C.S. (ed.). 1978. *Adaptive Environmental Assessment and Management*. John Wiley and Sons, London.

Holling, C.S. 1978. "Myths of Ecological Stability: Resilience and the Problems of Failure." in C.F. Smart and W.T. Stanbury (eds.). *Studies on Crisis Management*. Butterworth Publishers for the Institute for Research on Public Policy, Montreal.

Holling, C.S. 1986. "Resilience of Ecosystems: Local Surprise and Global Change." In W.C. Clark and R.E. Munn (eds.). *Sustainable Development of the Biosphere*. Cambridge University Press, Cambridge, pp. 292-317.

Holling, C.S. 1992. "New Science and New Investments for a Sustainable Biosphere."Arthur R. Marshall Jr. Laboratory of Ecological Sciences, Department of Zoology, University of Florida, Gainesville, FL, July 2. Prepared for the Biodiversity Project, International Institute of Ecological Economics and the Conference on Investing in Natural Capital—A Prerequisite for Sustainability.

Odum, Eugene. 1989. *Ecology and Our Endangered Life-Support Systems*. Sinnauer Associates, Sunderland, MA.

Odum, Howard. 1986. *Ecosystem Theory and Application*. Wiley, New York, NY.

Odum, Howard T. 1988. "Self-Organization, Transformity, and Information." *Science*, Vol. 242, November 25.

3 Industrial Ecology Methods and Tools

INTRODUCTION

Industrial ecology is a context for applying existing methods and tools, for adapting them to unique IE functions, and for creating new ones. Industrial metabolism was the earliest IE method to develop. Dynamic input-output models and design for environment are examples of older methods adapted for IE purposes.

Energy efficiency, pollution prevention, and life-cycle assessment are specific existing approaches that can gain strength from application in a larger IE system.

Full application of industrial ecology suggests many changes in the structures of organizations and information systems. So methods for organizational design and change management will be an important support to implementation processes.

To achieve their objectives, industrial ecologists will also need to develop new methods and tools and to build bridges with related fields such as ecological economics and sustainable business management.

Industrial ecology is a systemic context for applying existing methods and tools for designing and managing industry and for creating new ones. It provides connections across levels of design.

Levels of Design and Decision Making

Industrial ecology may be seen as a response to a nested set of challenges from individual product development to global policies. It can help guide design and management across these levels. *See transportation case illustrating three levels of design at the end of this chapter.*

Sustainable environmental performance requires design at the following levels:

- *Products*
- *Services*
- *Processes*
- *Materials and energy flows*
- *Facilities*
- *Business organizations/ missions/strategies*
- *Intercompany relations*
- *Community and regional infrastructure*
- *Societal institutions and policies.*

With this broad, multilevel context, the basic questions become:

How can we act with creativity and rigor to design effective environmental solutions at each of these levels?

How can we best evaluate alternative solutions?

How do we know the right level of design or management for approaching a particular issue?

How do we resolve conflicts across levels?

How do we maintain a coherent view of the whole system in order to manage design and management decisions well at any level?

Industrial ecology is seeking to create useful responses to this set of questions. It offers a whole systems context for effectively using existing tools and methods such as pollution prevention, energy efficiency, total quality environmental management, life-cycle analysis, and other valuable environmental management approaches. This integrative function may be the central contribution of industrial ecology to environmental management. Its proponents suggest that managers of the industrial system must treat it, at every level, as a set of organisms, subject to ecological constraints, like any other member of an ecosystem. Most other environmental management approaches fall short of this position. They define constraints in regulatory, economic, and engineering terms. These are necessary but not sufficient conditions for effective environmental management.

Industrial ecologists often state the field's objective as a closed-loop, resource efficient, and nonpolluting industrial system. Does industrial ecology actually have the methods and tools needed to achieve this? Several useful methods are now available for this task, but the IE toolkit still needs much development. Industrial ecologists are adapting existing methods and tools such as concurrent engineering and logistics engineering. They are also identifying new methods not now available.

Design Requirements for IE Methods and Tools

Industrial ecology's unique contribution is the context for applying existing methods, for redesigning them, or for developing new ones. This context calls for certain capabilities. Designers and managers need IE methods that enable them to:

- View the human system under consideration as part of its ecosystem and larger natural systems.
- Coordinate planning and action across time and space (local to global; short- to long-term). This implies connecting incremental changes into broader changes.
- Coordinate industrial redesign with emerging understanding of environmental challenges (global to local).
- Balance economic and environmental considerations (human needs and ecological needs and constraints).
- Balance efficiency and resilience in system design.

Industrial ecologists will also need a basic capability to integrate the application of any one method with the use of others needed for a given task.

Industrial ecologists are not trying to lay claim to specific methods such as energy efficiency or pollution prevention. They simply view IE as an effective organizing framework for applying many existing methods and tools, as well as for developing new ones.

An example of integrating applications of methods: Creating a regional strategy for sustainable development would require an inventory of the area's ecological conditions and constraints (ecology); a survey of the flows of materials and energy in human systems (industrial metabolism); methods for improving industrial, commercial, and household use of energy (energy efficiency) and materials (pollution prevention and recycling); and a means of assessing alternative strategies (dynamic input-output modeling and other simulation tools). The effort would also need input from urban planning, economic development, and citizen organizations.

INDUSTRIAL METABOLISM

Summary of the Field

Industrial metabolism (IM) traces material and energy flows from initial extraction of resources through industrial and consumer systems to the final disposal of wastes. Robert Ayres first developed this form of analysis in the 1970s, and it has become an important foundation of industrial ecology.

IM can be applied usefully at many different levels: globally, nationally, regionally, by industry, by company, and by site. A few companies have conducted environmental audits based on this method. *See the sidebar on Odwalla, Inc.* Regional application gives valuable insight into the sustainability of industry in natural units such as watersheds or atmospheric basins. Mapping sources, processes of transformation, and sinks in a region offer a systemic basis for public and corporate action. *See the sidebar on studies of heavy metals flows in the Rhine Basin.*

Industrial metabolism analysis highlights the dramatic difference between natural and industrial metabolic processes: in natural systems materials flow in closed loops with near universal recycling. Industrial systems are often very dissipative, leading to materials concentrations too low to provide value but high enough to pollute. Dissipative use is where materials are degraded, dispersed, and lost in the course of usage. In these terms, an inclusive definition of waste would be: *dissipative use of natural resources*. Any release to the environment in dissipative form (i.e., too dilute or chemically locked up to be of economic value) is nonsustainable, because it moves material "out of reach" of the industrial cycles that depend on it.

"Many materials uses are inherently dissipative. The materials are degraded, dispersed and lost in the course of a single normal usage . . . (including) food and fuels, packaging materials, lubricants, solvents, flocculants, antifreezes, detergents, soaps, bleaches and cleaning agents, dyes, paints and pigments, most paper, cosmetics, pharmaceuticals, fertilizers, pesticides, herbicides and germicides . . . Except for food and fuel, most such uses are nonessential in that technologies could be developed to eliminate the need for them."

(Ayres 1989B)

From a policy standpoint, the most important tracking is probably the concentration and chemical availability of the material. Materials that are lost in low concentration streams or low energy forms that have no economic value or that need large infusions of energy to be "replenished" are likely to be released to the environment and are intrinsically wasteful uses.

IM, Energy Flows, and Water Usage

Industrial metabolism studies have tended to focus on flows of materials. The method is also useful in analysis of energy and water flows. The U.S. energy system is a good illustration of how this approach can bring insights to business and government leaders working in this field. There are very low system efficiencies in terms of the services actually delivered to society (mobility, heat, light, commodities, etc.). These massive wastes define potential business opportunities and directions for policy development.

The total input of energy to transportation in the U.S. equals the total energy demand of the Japanese economy, 20 quadrillion BTUs (quads). *18.4 quads of this energy is wasted* in heat dissipated from engines, vehicles idling in gridlock, and at other points along the flow. This enormous level of waste suggests both fundamental redesign of products and of transportation systems (Basset 1996). *See the transportation cases at the end of this chapter and the scenario on a developing country at the end of the book.*

U.S. power plants also match the size of total Japanese energy *usage* in the waste heat they send up the stacks: 20 quads. Utilities are beginning to find ways to put this energy (and investment) to work through co-location with industrial plants that can use it. But even at the Kalundborg, Denmark, Eco-Industrial Park, the power plant manager feels he has just begun to tap this resource.

A full industrial metabolism study of a national energy system would also model its materials inputs and outputs, including all emissions.

An IM analysis of water usage in the U.S. shows a similar level of inefficiency and waste. Very little wastewater is directly recycled in the system, with most reuse being for irrigation. In addition, a significant amount of water remains unaccounted for or is lost in transit (through leaks in pipes, irrigation systems, etc.). The use of water in the energy sector almost equals that in agriculture. Overall, a metabolic view of the water and wastewater system shows high levels of throughput, significant waste, and little reuse of resource and waste streams.

Odwalla, Inc., Northern California's largest manufacturer and distributor of fresh juices, brought in Berkeley-based Gil Friend and Associates to conduct an eco-audit of the metabolism of its operations in 1994-95. The team charted resource flows (including energy, water, and commodities), procurement (including recycled content and agricultural production), waste generation, and recycling. The eco-audit suggested "key indicators" to track, including energy and water used and waste generated per gallon of product. These will be used in parallel with financial and production indicators, in the coming year.

The metabolic eco-audit at Odwalla's main production facility (Dinuba, CA) disclosed significant opportunities for efficiency gains in a new, already well-run, "just-in-time" facility. These included new potential profit centers from "waste" streams. The next step will be integrating the eco-audit process into company information systems to make information about environmental quality and efficiency progress readily accessible to all employees throughout the year.

Guidelines for Improving the Metabolic Pathways of Industrial Processes and Materials Use

- Reduce dissipative uses of materials through change in product or process design, enhancement of reuse, and recycling.
- Change product design to eliminate toxic materials from being dissipated into the environment as a function of use.
- Reduce the number of steps in processes to achieve greater energy and resource efficiencies.
- Create on-site, on-demand production of any required toxic materials.
- Emulate biological metabolism in temperature and pressure and in cyclic processes.

In Lower Saxony, Germany, three medium-sized companies have become demonstration sites for development of ecological controls, parallel to their financial control systems. (They are manufacturers of foil packaging, wallpaper, and paint industry supplies.) They have conducted an ecological balance study, identifying all material and energy flows. This study goes beyond plant boundaries to cover the entire product life cycle.

The companies subjected the data to a "weakness analysis," which helps them set environmental objectives and assess alternative products, technologies, processing materials, and production methods. The results have enabled them to discover ways to conserve raw materials, energy, and water and realize cost savings.

(Griefahn 1994)

- Achieve overall system efficiencies as a cooperative effort between suppliers and customers.
- Develop metrics for assessing an industrial system's efficiency and productivity, for example:
  1. *The ratio of virgin to recycled materials:* This ratio could be used in assessing a product, production system, company, industry, or economy. The lower the relative draw upon virgin materials (to replace materials lost from dissipation) the closer the system would be to sustainable.
  2. *Ratio of actual/potential recycled materials:* This ratio between volume of materials that could be recycled to the fraction actually recycled would also be useful at levels ranging from a production line to an economy. Ayres suggests a proxy measure could be the ratio of secondary supply to total supply of final materials.
  3. *Ratio of renewable/fossil fuel sources:* This ratio would tend to remain fairly stable for older plants but would be useful in designing new facilities, in reconstruction of plants, and at a corporate level.
  4. *Materials productivity:* The economic output per unit of material input.
  5. *Energy productivity:* The economic output per unit of energy input.
  6. *Resource input per unit of end-user service:* This ratio assesses resource use against the useful function gained and maintained for the end-user. It is suggested by Walter Stahel's work envisioning highly durable products delivering function to the customers of service-oriented companies. *See the next chapter.*

Materials and energy productivity, indicators 4 and 5, are fundamental IM measures, useful for assessing resource efficiency at a plant, company, industry, or county level. Monitoring indicator 6 would enable a product manager, a company, or an industry to plan achievement of even higher levels of efficiency.

Ratios 1, 2, and 3 are from Ayres and Simonis, 1994, Chapter 1.

Process for Applying IM

The generic steps in conducting an industrial metabolism analysis are provided below. The level of detail and specific elements researched would depend upon the system in focus.

The basic task is to build a systems model of materials and energy flows through the domain under consideration. The study might focus on materials or energy needs that are diminishing natural resources, ones in which levels of waste appear to be unnecessarily high.

1. Map materials and energy flows.
   - ◆ Define system and subsystem boundaries to determine where materials actually enter or leave the system in focus. (This system definition is both spatial and temporal.)
   - ◆ Identify sources of all energy and materials, energy costs of extraction and processing, and emissions generated.
   - ◆ Inventory all process and product materials.
   - ◆ Identify sources of emissions for materials being studied. Include all point and diffuse sources along the cycle from extraction through refining, manufacturing, and use to disposal.
   - ◆ Estimate quantities of emissions to air, water, and soil for each source.

2. Analyze critical ratios.
   - ◆ Calculate virgin/recycled materials ratio.
   - ◆ Calculate percent of energy and materials going into product, into recycling (internal or external), into emissions, and into dumps.
   - ◆ Estimate total cost of producing wastes and emissions, hidden and explicit.

3. Develop scenarios for improving performance in the context of this whole systems analysis. Determine most effective program for change and set targets.

An IM study of heavy metals pollution in a plant and in the larger region surrounding the facility could discover that the plant's controls and recycling have reduced its point sources to a low level compared with the non-point sources distributed through the region. The study would pinpoint areas where plant performance could still be improved. But priority for public policy making would be developing an effective program for dealing with the non-point sources through community education, infrastructure, management of government services, and the like.

An internal IM study might explore an issue like the balance between virgin and recycled materials use, developing a model of materials use, industrial processes, and both environmental and economic costs/benefits. This would enable decision makers to evaluate scenarios for change to decrease use of virgin materials.

A specific application of this process would be an audit of a plant's or company's energy and materials flows. Or, the companies in an industrial park could combine data from their individual audits to discover the potential for exchange of energy and materials by-products. A private/public partnership could do an IM study of toxic materials flows in a region to define where in the system interventions would generate the greatest reductions in pollution. *Researchers have conducted such a study of heavy metals flows in the Rhine Basin.* *See the sidebar on page 46.*

Benefits of IM

IM's integrated view of the interactions between energy, materials, and the environment helps managers and regulators avoid narrowly conceived or short-run "quick fix" policies, which may actually have negative impacts. It supports managers in identifying and evaluating potential opportunities for cost savings and improvements in environmental performance. IM provides a logical, semi-quantitative and disciplined means of assessing sustainability within a specific unit of the economy, such as an individual plant, firm, or municipal facility. Industrial metabolism enables users to better determine the full costs of materials, factoring in the value of nonrenewable resources and environmental pollution. For a company, analysis of its industrial metabolism as a whole or at specific facilities provides a holistic view to guide internal policies and interactions with regulators and other environmental stakeholders. Industrial metabolism studies could play a vital role in regional economic development planning, in both developed and developing countries.

Challenges/Obstacles to IM

- Data needed for full analysis may not be available. (IM often requires extrapolation from existing data.)
- Processes for applying IM need more development for use in corporate and plant settings.
- Many waste materials are unusable *in the quantities generated*, including nitric acid, sulfur oxides, lignin wastes, fly-ash from coal, etc. Accelerated R&D is required to develop uses, processes to make them usable, or substitutes.

- Present uses of many materials are inherently dissipative (the materials are degraded, dispersed, and lost in the course of a single normal usage). IM is useful to highlight this fact, but solutions will be found only through redesign of products or alternative ways of satisfying the user's need.

Resources on Industrial Metabolism

This chapter was written with important contributions from Robert Scott Butner, Battelle Seattle Research Center, Seattle, WA.

The International Institute for Applied Systems Analysis (IIASA) has been the site for IM work by Robert Ayres and other researchers. IIASA's address is A-2361 Laxenburg, Austria. Telephone: 43-02236-71521-0, Fax: 71313, Internet: postmaster@iiasa.ac.at. The World Wide Web address is http://www.iiasa.ac.at/ Information on the Rhine Basin IM study is available online here.

Robert Ayres is on the faculty at the European Institute of Business Administration (INSEAD) near Paris. INSEAD, Blvd. de Constance, 77305 Fontainebleu, Cedex France. Telephone: 33-1-6072-4000.

Allen, David T., and Behmanesh, Nasrin. 1994. "Wastes as Raw Materials." In Braden R. Allenby and Deanna J. Richards (eds.) *The Greening of Industrial Ecosystems*. National Academy Press, Washington, DC.

Ayres, Robert U. 1989A. "Industrial Metabolism." In Ausubel, Jesse H. and Sladovich, Hedy E. *Technology and Environment.* Washington, DC: National Academy Press, pp. 23-49.

Ayres, Robert U. 1989B. "Externalities: Economics & Thermodynamics." In Archibugi and Nijkamp (eds.). *Ecology and Economy: Towards Sustainable Development*. Kulwer Academic, Netherlands.

Ayres, Robert U. 1994. "Industrial Metabolism: Theory and Policy." in Allenby, Braden R., Richards, Deanna (eds.). *The Greening of Industrial Ecosystems*. Washington, DC. Available through National Academy of Engineering Press Office 202-334-3313 or 1-800-642-6242).

Ayres, Robert U., et al. 1989. *Industrial Metabolism, the Environment, and Application of Materials-Balance Principles for Selected Chemicals*. A research report of the International Institute for Applied Systems Analysis, October.

Ayres, Robert. U., and Simonis, Udo E. (eds.). 1994. *Industrial Metabolism—Restructuring for Sustainable Development*. UN University Press, Tokyo.

Basset, David. 1996. "Quantified Energy Flow Charts Developed for US-EPA Futures Unit." In paper by David Rejeski published in Deanna J. Richards (ed.), *The Industrial Green Game: Implications for Environmental Design and Management*. National Academy Press, Washington, DC.

Brodyanski, V., Sorin, M., and Le Goff, P. 1994. *The Efficiency of Industrial Processes: Energy Analysis and Optimization*. Elsevier, New York. Energy is a measure of the availability of energy, an important metric for industrial metabolism analysis.

Eisenhauer, J., and Cordes, R. 1992. *Industrial Waste Databases, Hazardous Waste and Hazardous Materials*, 9 (1) 1-19. (The entire issue is dedicated to industrial waste generation and management.)

Griefahn, Monika. 1994. "Initiatives in Lower Saxony to Link Ecology to Economy." In Socolow, Robert, et al., *Industrial Ecology and Global Change*. Cambridge University Press, NY, pp. 426-8.

Stigliani, W.M., and Anderberg, S. 1991. "Industrial Metabolism and the Rhine Basin." *Options*. International Institute for Applied Systems Analysis, Laxenberg, Austria, September.

The International Institute for Applied Systems Analysis (IIASA) has completed the first phase of an industrial metabolism study of the Rhine Basin, the largest application of IM so far. This basin is probably the most heavily industrialized region in the world.

The study examined sources of pollution and pathways by which pollutants end up in the river for the whole basin. Materials studied include cadmium, lead, zinc, lindane, PCBs, nitrogen, and phosphorous.

The results suggest that in the Rhine Basin industry has made major progress on reducing emissions. However, there are increasing flows of pollution from "on-point" diffuse sources, including farms, consumers, runoff from roads and highways, and disposal sites. These findings are of great value in designing policy, industrial practice, and public education.

A second phase of the IIASA study will continue research on the Rhine and also include Upper Silesia in the Upper Elbe/Oder Basin. The focus of this study is the relationship among heavy metals mobilization, acidification, and land use.

DYNAMIC INPUT-OUTPUT MODELING

Summary of the Field

Faye Duchin, former Director of New York University's Institute of Economic Analysis, has created "what if" tools based upon the foundation of industrial metabolism and structural economics. *See the sidebars on the next three pages.* These dynamic input-output (IO) models enable business and policy decision makers to perceive the broad business, economic, and environmental implications of systemic technical change. The Nobel prize work of Wassily Leontief on input-output models of exchanges among industries is Duchin's foundation in economics.

The IO models add environmental resource accounts to economic information about the 100+ industrial sectors found in standard national input-output tables. By incorporating a time dimension, Duchin has created a means of analyzing the total impacts of alternative scenarios of industrial change: How would the changes affect the environment, businesses in the target industry, and their major suppliers and customers?

Duchin's work provides "an analytic framework for considering the economic implications of complex webs of technical changes. . . . Dynamic input-output models are used to develop a set of possible solutions rather than a single optimal one. . . [making it] possible to experiment with changes in input structures that might reduce water usage in production, for instance, or recover products of economic value . . . . A more complex set of results, involving economic and environmental trade-offs, can be evaluated" (Duchin 1992).

Duchin has applied IO modeling to issues of household consumption, an important first in IE. Most industrial ecologists focus on manufacturing.

"Structural economics emphasizes the representation of stocks and flows, measured in physical units, as well as associated costs and prices. . . . The variables representing such physical measures (like tons of steel or tons of carbon emissions), unlike variables that are essentially symbolic or index numbers, provide a direct link to technology and to the physical world with which industrial ecology is concerned. . . . The data are developed using technical expertise and practical experience as well as experimental results, technical records, and accounting information."

(Duchin 1992)

Process of Applying Dynamic Input-Output Modeling: A Corporate Level Application

A corporation evaluating its overall strategy for achieving sustainability or developing new products would

- Evaluate alternative scenarios for combined technological-organizational change in economic and environmental terms
- Set up an IO framework for continuing evaluation and feedback as the changes are instituted
- Determine impact on competitive situation
- Determine what policy initiatives to launch to enhance public climate and regulatory structure to encompass its innovative strategies.

An automobile manufacturer, for instance, might choose to study the impact on the environment and its own future of possible socio/technological changes such as

- Innovations in engine design resulting from much higher standards for emissions and fuel efficiency
- Systemic redesign of small vehicles as proposed by Amory Lovins

  See the RMI Hypercar case at the end of this chapter.
- An increase in U.S. fuel prices to the global average

- A dramatic increase in short to mid-distance rail transport and a resulting increase in demand for rolling stock and feeder motor vehicles.

In the IO study the auto manufacturer could build alternative scenarios such as

- Remaining focused on traditional motor vehicle transport through technological innovation needed to meet the regulatory and economic changes
- Developing and marketing lines of alternative vehicles (electric and hybrid-electric)
- Possible diversification into railcar production through acquisition of a current manufacturer and retooling some of the company's auto parts plants.

"Our models are 'open,' making use of exogenous information such as technological projections that are provided by engineers and other technical experts rather than being derived through the use of mathematical equations that describe idealized economic mechanisms."

(Duchin 1992)

Researchers would then go through these steps:

- Create conceptual models to develop the most useful research questions and to guide next steps.
- Build a data base of relevant data in a form the dynamic IO models can use, including, as appropriate,

    National accounts with industries selected for the study (if working in a model of the national economy)

    Environmental accounts reflecting resources and sinks (as well as wastes and emissions) needed to analyze the environmental impact of the technological changes in question

    The company's financial information, especially capital stocks, investments, etc.

    Data on capacity utilization and costs, stocks and flows for energy and materials

    Information on the technologies being evaluated, including projections of technical data for the future.

- Use existing strategic and technology innovation plans to develop detailed scenarios about alternative future paths.
- Evaluate each scenario from economic and environmental perspectives using the dynamic IO model.

The final products for the manufacturer would be a set of scenarios with assessment of the impact of each on corporate profitability, corporate stakeholders, and the environment. It would have a rationale to guide policy and public relations work around its decision. The IO modeling tool developed for transportation would continue to be useful for evaluation of new strategies as other environmental, technical, and social changes emerge.

Transportation From the Viewpoint of a Developing Economy

The transportation and economic development ministries of a developing economy might use IO modeling to evaluate alternative scenarios for creation of a transportation infrastructure and industry. Scenarios explored might include

- An auto- and truck-based highway system
- A rail-based, intermodal system
- Moderation of the need for travel through application of information technologies.

Some key elements in the model would include

- Vehicle efficiency and fuel use
- Emission characteristics and air pollution
- Demands on energy and material resources
- Economic and environmental implications of new roads, rail lines, telecommunications, and other infrastructure
- Congestion and travel times
- Choice among material processing technologies and the associated demand for material and energy resources
- Labor requirements and the capacity of the educational system
- Information system requirements.

In the mid-90s, Chinese leadership is projecting an industrial development strategy based on automobile and truck manufacturing and infrastructure. An IO study as outlined here could open an effective process for exploring alternative strategies.

A two-phase IO study by Duchin in 1994-95 focused on the use, disposal, and recycling of plastics in the U.S. This work is of particular importance to policy makers, given the volume of plastics in municipal solid waste flows (24 percent), the limits on future fill capacity, and the relatively low rate of recycling (4 percent). Duchin analyzed potential industrial consumption of the various forms of recycled plastics, finding that an optimistic forecast would still see 89 percent of the materials going to landfills by 2005. Source reductions of plastic use in products and packaging are likely to play a smaller role in reducing the overall waste stream than recycling. The study also analyzes policies creating obstacles to recycling and those facilitating the practice. The first phase of research was one of the few industrial ecology studies of household consumption and disposal patterns.

An IO model of the Brundtland Report

Dr. Duchin's application of IO modeling to the Brundtland report recommendations indicates that this landmark in sustainable development seriously underestimated the need for change.

"Our results show that if moderate economic development objectives are achieved in the developing countries over the next several decades, the geographic locus of emissions will continue its historic shift from the rich to the poor economies while total emissions of the principal global pollutants will increase significantly. This is true even under optimistic assumptions about pollution reduction and controls through the accelerated adoption of modern, commercially proven technologies in both the rich and poor countries." (Duchin 1993)

This IO analysis proposes that industrial ecology methods will be needed to achieve dramatic reductions in use of fossil fuels and near-zero emissions of persistent toxic chemicals. Duchin says IE will have to be applied in agriculture, forestry, fisheries, and other sectors beyond manufacturing to achieve a sustainable world.

(Duchin and Lange, 1994A, is a book providing a full report of this analysis of *The Brundtland Report, Our Common Future.)*

Benefits of IO Analysis

IO analysis enables users to consider the economic and environmental implications of complex webs of technical change (interactions among the ecosystem itself, systemic technical change, internal and external accounting, market forces, regulations, and international treaties). It serves as the basis for the development of incentive schemes, legislation, and international agreements. IO also can help identify bottlenecks in research and development that will not be resolved in a timely fashion by private markets. Such models can evaluate the costs and potential contribution to reducing pollution of alternative design for environment strategies. Finally, IO models provide the framework for evaluating alternative scenarios in an economy-wide context.

Challenges of IO Analysis

A significant time lag must be factored in since the IO models depend on aggregations of data from Department of Commerce, the census, and other national agencies. Building a model is a lengthy process requiring assembly of diverse data bases (economic, environmental, and technical). Some required data may be incomplete, requiring estimates. A powerful global IO model and ones for the U.S. and other countries are already available, needing additional input relating only to particular technical innovations to be explored. The creation of a model also requires educating the stakeholders using it. Participants in a modeling exercise need to understand that the process is designed to guide their search for solutions, not to automatically provide the answers.

Resources on IO Analysis

Faye Duchin is Dean, School of Humanities and Social Sciences, Rensselaer Polytechnic Institute, 110 8th St., Troy, NY 12180. Telephone: 518-276-6575, Fax: 518-276-4871, Internet: duchin@rpi.edu.

Glenn-Marie Lange is Director, Institute for Economic Analysis, New York University, 269 Mercer St., New York, NY 10003. Telephone: 212-998-7980, Fax: 212-998-7484.

Duchin, Faye. 1991. "Prospects for Environmentally Sound Economic Development in the North, in the South, and in North-South Economic Relations: The Role for Action-Oriented Analysis." In *Journal of Clean Technology and Environmental Sciences*. Vol. 1, No. 3/4, pp. 225-238.

Duchin, Faye. 1992. *"Industrial Input-Output Analysis: Implications for Industrial Ecology."* In *Proceedings of the National Academy of Sciences of the USA*, Washington, DC, Vol. 89, No. 3, pp. 851-855.

Duchin, Faye. 1993. "Input-Output Analysis and Industrial Ecology." In Allenby, Braden R., and Richards, Deanna (eds.). *Greening Industrial Ecosystems*. Washington, DC. Available through National Academy Press Office (202-334-3313 or 1-800-642-6242).

Duchin, Faye. 1996. "Ecological Economics: The Second Stage." In Costanza, R., et al. (eds.). *Down to Earth: Practical Applications of Ecological Economics*. Island Press, Covelo, CA.

Duchin, Faye, and Lange, Glenn-Marie. 1993. "Development and the Environment in Indonesia: An Input-Output Approach. Final Report." Institute for Economic Analysis, NYU, NY.

Duchin, Faye, and Lange, Glenn-Marie. 1994A. *The Future of the Environment: Ecological Economics and Technological Change*, Oxford University Press. In U.S. 1-800-451-7556. This book is a detailed IO analysis of the Brundtland Report, *Our Common Future*. The authors' analysis challenges central assumptions of this landmark work on sustainable development.

Duchin, Faye, and Lange, Glenn-Marie. 1994B. "Household Use and Disposal of Plastics: an Input-Output Case Study for New York City." Report to the AT&T Industrial Ecology Faculty Fellowship Program. Institute for Economic Analysis, NYU, NY.

Duchin, Faye, and Lange, Glenn-Marie. 1994C. "Strategies for Environmentally Sound Economic Development." In *Investing in Natural Capital: The Ecological Economic Perspective*. A. Manson, C. Folke, R. Costanza, and M. Hammer (eds.). Covelo, CA: Island Press.

Duchin, Faye, and Lange, Glenn-Marie. 1995A. "The Use, Disposal, and Recycling of Plastics in the U.S." Report to the AT&T Industrial Ecology Faculty Fellowship Program. Institute for Economic Analysis, NYU, NY.

Duchin, Faye, and Lange, Glenn-Marie. 1995B. "Households and Eco-Restructuring: The Social Dimension of Ecological and Development Economics. Institute for Economic Analysis, NYU, NY.

From 1991-93 , the Indonesian government worked with Duchin and her staff in modeling development alternatives for the country. They built three basic technological scenarios and analyzed them with IO models predicting moderate and high growth rates for each. One scenario assumed continuation of present moderate trends in increased energy and materials efficiencies; a second assumed more aggressive government measures for resource efficiency and environmental protection; and a third added measures for cleaner and more energy-efficient energy production.

The team analyzed potential economic and environmental impacts of these scenarios in terms of changes in 15 major industrial sectors, including agriculture and forestry, food processing, pulp and paper, chemicals, iron, and steel. Environmental impacts included land use and land degradation, local and global air pollution from fossil fuel energy production, water withdrawals, and pollution.

The project was designed to train local staff in the process of dynamic IO modeling and install a national model for continuing use by public and private sector stakeholders. This will enable Indonesians to integrate environmental priorities into an effective development planning process.

LIFE-CYCLE ASSESSMENT

Summary of the Field

Life-cycle assessment (LCA) is a method that flows naturally from the questions posed by industrial metabolism. One phase of LCA—improvement assessment— is the context for development of the allied tool, design for environment (DFE). LCA tools focus on quantifying the environmental burdens of a product, process, or activity, looking at the whole cycle from extraction of resources through recycling or disposal.

A few examples of LCA studies include assessing the relative merits of paper or plastic grocery bags, comparing the total environmental impacts of electric vs. conventional automobiles, and a Proctor & Gamble study of impacts of household cleaning products that led to development of new cold water products. LCA is developing in parallel with industrial ecology, and there is a rich literature available on it. We will offer only a brief summary and resource list here.

Components of LCA

After an initial definition of the goals and scope of the assessment, an LCA team may conduct three types of analysis:

1. Inventory analysis: identification and quantification of energy and resource use and environmental releases to air, water, and land
2. Impact analysis: technical qualitative and quantitative description and assessment of the consequences for the environment
3. Improvement analysis: evaluation and implementation of opportunities to reduce environmental burdens (Keoleian 1994).

Research and specific tool development so far have focused primarily on the first type of analysis. Checklists, templates for flow diagrams, and tools for conversion of data from one type to another are all under development.

Benefits of LCA

Life-cycle assessment enables product or process design teams to evaluate all environmental impacts of components and activities. For instance, LCA of an electric car includes analysis of the emissions from generating the energy to charge its batteries. The outcomes point to changes in design that can reduce net resource demand and emissions. External public groups can use LCA to evaluate environmental claims for competing products.

"The life-cycle assessment is an objective process to evaluate the environmental burdens associated with a product, process, or activity by identifying and quantifying energy and material usage and environmental releases, to assess the impact of those energy and material uses and releases on the environment, and to evaluate and implement opportunities to effect environmental improvements. The assessment includes the entire life cycle of the product, process, or activity, encompassing extracting and processing raw materials; manufacturing, transportation, and distribution; use/reuse/maintenance; recycling; and final disposal."

(SETAC 1993)

Challenges of LCA

LCAs require extensive data gathering and analysis and can be fairly expensive. There are many gaps in the data now available, requiring approximation and extrapolation. The environmental impacts of many substances are not well understood. Interactions with the environment are often quite complex. Further, the initial definition of the goal and scope (boundaries of the system under analysis) is essentially a subjective step.

Researchers at the University of Michigan applied life-cycle design framework in a research project with Optical Imaging Systems (OIS). OIS is a U.S. manufacturer of high-performance, active-matrix liquid crystal displays, one of the leading flat panel display technologies. The study evaluated OIS's environmental management system and how environmental performance may impact competition in the industry. Metrics were developed to measure environmental performance in a factory simulation model. Strategies for improvement were recommended according to incremental, reengineering, and future approaches.

(Keoleian et al. 1994)

LCA Resources

Battelle Life Cycle Management web page: http://www.estd.battelle.org/sehsm/lca/.

Society for Environmental Toxicology and Chemistry, LCA web page address: http://www.setac.org/lca.html.

Volvo web page on LCA and enviromental management: http://www.car.volvo.se/environment/management/eps.html.

Keoleian, Gregory A. (ed.). 1994. *Product Life-Cycle Assessment to Reduce Health Risks and Environmental Impacts*. Noyes Press. Park Ridge, NJ.

Keoleian, Gregory A. (ed.). 1997. *Industrial Ecology of the Automobile : A Life Cycle Perspective.* Society of Automotive Engineers.

Keoleian, Gregory A., Glantschnig, Werner J., and McCann, William. 1994. "Life Cycle Design: AT&T Demonstration Project." *Proceedings of IEEE International Symposium on Electronics and Environment*. San Francisco, 2 May 1994. Piscataway, NJ: IEEE Service Center, 1994.

Keoleian, Gregory A., and Menerey, Dan. 1993. *Life Cycle Design Guidance Manual*. US-EPA Risk Reduction Engineering Laboratory, Cincinnati, OH, Publication number EPA/600/R-92/226.

SETAC, Society of Environmental Toxicologists and Chemists. 1993. "Guidelines for Life-Cycle Assessment: A Code of Practice." Pensacola, FL.

Vignon, B.W., et al. 1993. "Life-Cycle Assessment: Inventory Guidelines and Principles." U.S. EPA Risk Reduction Engineering Laboratory, Cincinnati, OH, Publication number EPA/600/R-92/036.

DESIGN FOR ENVIRONMENT

Summary of the Field

> In the short term, design for environment (DFE) is the means by which the still vague precepts of industrial ecology can in fact begin to be implemented in the real world today. DFE requires that environmental objectives and constraints be driven into process and product design and materials and technology choices (Allenby 1994).

DFE is a systemic approach to decision support for designers, developed within the industrial ecology framework. DFE teams apply this approach to all potential environmental implications of a product or process being designed—energy and materials used; manufacture and packaging; transportation; consumer use, reuse, or recycling; and disposal. Initial application has been in the electronics and chemical industries.

The tools enable consideration of these implications at every step of the production process from chemical design, process engineering, procurement practices, and end-product specification to post-use recycling or disposal. DFE also enables designers to consider traditional design issues of cost, quality, manufacturing process, and efficiency as part of the same decision system (Graedel and Allenby 1995).

Design for environment is an excellent example of adapting earlier methods and tools in an industrial ecology context. Its sources are life-cycle analysis, design for X, and concurrent engineering. DFE focuses on the LCA stage of improvement assessment, enabling design teams to weigh options for improvement of environmental performance while attending to the traditional design issues highlighted by design for X (DFX) systems. Pollution prevention teams could make effective use of DFE. Industrial metabolism analysis would be a valuable context for DFE based design.

"Design for Environment designates a practice by which environmental considerations are integrated into product and process engineering design procedures. Accordingly, DFE is an attempt to implement industrial ecology principles into a systems analysis approach to environmental management. DFE practices require consideration of all potential environmental implications of the product or process being designed, not just those that are mandated by law. DFE practices are meant to develop environmentally compatible products and processes while maintaining product price/performance and quality standards."

(Allenby and Fullerton 1991-92)

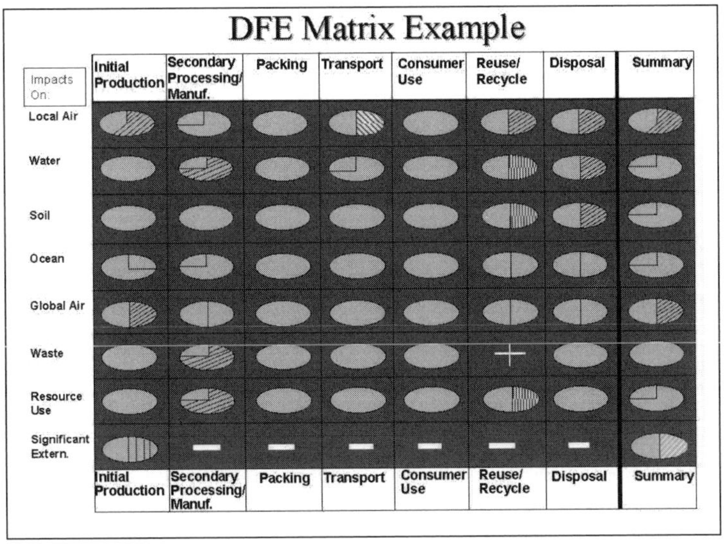

DFE Matrix Example

Impacts On:	Initial Production	Secondary Processing/ Manuf.	Packing	Transport	Consumer Use	Reuse/ Recycle	Disposal	Summary
Local Air								
Water								
Soil								
Ocean								
Global Air								
Waste								
Resource Use								
Significant Extern.								
	Initial Production	Secondary Processing/ Manuf.	Packing	Transport	Consumer Use	Reuse/ Recycle	Disposal	Summary

Two AT&T executives, Braden Allenby and Thomas Graedel, have devoted a major portion of their industrial ecology text (Graedel and Allenby 1995) to discussion of DFE and LCA. In contrast to LCA, they recommend a largely qualitative rather than quantitative approach in DFE. They believe the design task is often too complex to lend itself to quantitative analysis. In complex design situations, they state, "Quantitative models simply eliminate too much information that could be valuable to the designer in reaching design decisions. In addition, too many value-judgments are buried in the data; and the data itself is too incomplete to drive a quantitative system."

Allenby and Graedel offer tools to help designers compare alternative options in a more systemic way and to graphically demonstrate those aspects of design that would most improve the environmental performance of the system in question. DFE matrices have question sets for each cell that designers use to score each activity against environmental concerns. Typically DFE uses more detailed matrices to feed evaluations into each area of a more general analysis. The matrices seek to provide a design team (speaking many different professional languages) with a common framework for seeing the whole project and the place of each part in the whole.

Materials choices in automobile design illustrate the sort of complex systems issue DFE can address: iron as a basic material in cars involves a burden on natural resources and the greatest weight, but it is recycled at a very high rate. Plastics lighten automobile weight, reducing fuel demand, but are now difficult to recycle. Therefore, they generate much more solid waste than iron. Aluminum manufactured from ore requires a large energy investment out front, but its light weight reduces fuel consumption and it is readily recycled.

Making design judgments about these trade-offs requires a sophisticated system, balancing the burden on virgin resources, energy usage at each stage of the product life cycle, recyclability of materials, manufacturability, and costs.

DFE Activities and Tools

Allenby and Graedel identify two groups of DFE activities within the firm. *Generic* DFE involves the broad programs for improvement of environmental performance across all design functions. These include development of "green" accounting systems, standard components lists, and supplier and internal specifications and standards. *Specific* DFE activities, on the other hand, integrate this approach into product realization processes to evaluate options for a particular product, process, or input. Design teams expand DFX software and checklists to include environmental issues along with design for manufacturability, testability, safety, etc.

Allenby describes two DFE design tools:

1. The DFE template is a generic set of procedures and practices that can be modified to match design practices and requirements of a specific firm. This tool is compatible with existing Design for X systems and existing design practices. A set of matrices enables designers to graphically rank specific concerns across toxicity/exposure, environmental, manufacturing, and social/political questions. The graphic conventions also allow ranking the level of certainty in available information.

2. The Design for Environment Information System (DEIS) summarizes relevant environmental health and safety, social, economic, and regulatory data applicable to specific design options. (This is a data base unique to each company.) Ideally, the DEIS will be
   - comprehensive, providing sufficient data for a balanced, systemic evaluation of a product or process design
   - multidimensional in approach
   - nonprescriptive, providing context for informed decision-making and trade-offs
   - basically qualitative, not quantitative
   - indicative of the degree of uncertainty associated with its data (Allenby and Fullerton 1991-92).

"Limited experience indicates DFE provides firms with product cost advantages while reducing uncertainty and risk to product lines generated by changes in environmental statutes and regulations."

(Allenby and Fullerton 1991-92)

Design for environment integrates work across many functions: marketing, R&D, manufacturing, quality, and procurement. DFE also calls for greater integration of personnel from suppliers in the design process. Implementing it requires good cross-organizational design. Executive level buy-in is required to achieve the combined organizational, financial, and technical changes implied by DFE.

Benefits of DFE

DFE brings environmental and financial design considerations together in a whole systems framework. It frames the environmental design issues for the industrial designer or process engineer in the same way that manufacturability, cost competitiveness, and quality now do.

DFE's systems approach could be an effective way of integrating the multiple issues designers must consider while integrating well with concurrent engineering practices already in place in many companies.

Challenges of DFE

Incremental improvement in a product through DFE may mask a broader need to simply not continue making that product.

We lack a sufficient industrial materials data base providing reliable and accessible information on environmental impacts of many materials, chemicals, and processes.

Full application of DFE demands involvement of the public sector in defining values for the design trade-off process involved. Industry alone cannot be responsible for weighing the relative value of one material's health effects (i.e., lead in solder) and another material's relative scarcity (indium as a substitute).

DFE is largely qualitative rather than quantitative analysis (by intention).

While design for environment integrates work across many functions: marketing, R&D, manufacturing, quality, and procurement, some organizations do not support this cross-functional approach.

Designers themselves may feel besieged by too many design requirements. So many considerations have been added in a piecemeal fashion to their task that they feel overloaded.

There is much divergence and controversy concerning environmental objectives, with regulators setting some targets, scientists in different environmental realms proposing others, and environmental organizations calling for still others. Design for environment needs to be grounded in a more effective public/private dialogue in order to sort through these conflicting objectives.

DFE Resources

Design for Environment at Stanford http://dfe.stanford.edu/. This web site includes links to many other DFE and green design research home pages in the U.S. and overseas.

Design for Environment at DOE: http://w3.pnl.gov/2080/DFE/home.html.

Georgia Tech, Environmentally Conscious Design and Manufacture Research in the Systems Realization Laboratory: http://srl.marc.gatech.edu/research/ECDM-page.html.

Massachusetts Institute of Technology web site for Design for Environment Project: http://web.mit.edu/afs/athena/org/c/ctpid/www/tbe/dfemain.html.

Rochester Institute of Technology, the Environmentally Conscious Manufacturing Program (ECMP): http://www.isc.rit.edu/~633www/research/ecm/ecm_center.html. The Rochester site includes information on design for remanufacturing.

Allenby, Braden R. 1994. Integrating Environment and Technology: Design for Environment. In Allenby, Braden R., Richards, Deanna (eds.). *Greening Industrial Ecosystems*. Washington, DC: National Academy Press Office.

Allenby, Braden R., and Fullerton, Ann. 1991-1992. "Design for Environment—A New Strategy for Environmental Management." *Pollution Prevention Review*, Winter.

Allenby, Braden, and Graedel, Thomas. 1995. "Defining the Environmentally Responsible Facility." Paper from 1994 Woods Hole Workshop on Industrial Ecology. Applying DFE to industrial facility design. See also Allenby and Graedel 1995A. pp. 285-289.

Fiksel, Joseph. 1996. *Design for Environment: Creating Eco-Efficient Products and Processes*. Mc-Graw Hill, New York. (This excellent book provides a broad overview of DFE, including a DFE framework, case studies, and guidance for taking the next steps.)

Graedel, T. E., and Allenby, B. R. 1995. *Industrial Ecology*. Prentice Hall, Englewood Cliffs, NJ. (This first university textbook on IE includes extensive discussion of design for environment and life-cycle assessment.)

Lenox, Michael and John Ehrenfeld. 1995. "DFE: A New Strategic Framework." Total Quality Environmental Management. Summer.

POLLUTION PREVENTION (P2)

Summary of the Field

Pollution prevention (P2) is a well-developed field of environmental management that focuses particularly on the design of industrial processes within plants to reduce and eliminate wastes. This approach has led to development of many strategies, assessment methods, and a wide range of "clean technologies" that often improve both environmental and economic performance.

P2 does not generally address the relationships between plants. Nevertheless, it is complementary to industrial ecology strategies for improving performance among sets of plants. At present there are regulatory constraints on the intercompany exchange of waste materials, a primary IE strategy. Design for environment tools could be very useful to pollution prevention teams, providing a systemic means for weighing trade-offs among competing options for change.

To guide companies in identifying pollution prevention opportunities, the EPA created an environmental management hierarchy (US-EPA 1992A). The hierarchy builds environmental protection into the industrial waste management process. It places highest priority on reduction of hazardous wastes at the source and internal recycling rather than treatment and/or disposal of wastes to land, air, and water. The strategies, from most to least favorable, are

- Source reduction
- Recycling
- Treatment
- Disposal.

The preferred P2 strategy, source reduction, eliminates or reduces the use of hazardous materials and/or the generation of hazardous waste at the point of generation. It includes the following approaches:

- Materials substitution
- Product substitution
- Product reformulation
- Process or equipment changes

◆ Improved plant operations (housekeeping, material inventory control, spill prevention, etc.).

According to EPA guidelines, off-site recycling, although generally a good waste management technique, is not pollution prevention. This limitation may not always be supportive of the concept of industrial ecosystems. When transporting materials off site for recycling, companies increase the risk of releases during transport. Off-site recycling also increases worker exposure to chemicals because of an increase in material handling. In addition, residuals generated from the recycling process must often be treated, transported, and disposed of.

Some of EPA's concerns regarding off-site recycling may be met in an industrial ecosystem approach. By clustering industries in close proximity, transportation hazards are reduced, lowering risk to drivers, the public, and the environment. Thus, "off-site recycling" within an industrial park more closely resembles on-site recycling in terms of risk and benefit.

At a Ciba-Geigy plant in New Jersey, two improvements in the dye-making process made possible a 40 percent increase in yield, reduced iron waste by 100 percent, and reduced total organic carbon waste by 80 percent, resulting in an annual cost savings of $740,000. Additional analysis showed that the process improvement could increase the yield another 15 percent.

(Romm 1994)

A related issue is waste and material exchange between companies. Waste exchange, although not an official P2 strategy, is generally encouraged by EPA as an alternative to treatment and disposal after source reduction opportunities have been exhausted (US EPA 1994). However, many regulatory barriers exist that discourage material and waste exchange. Essentially, complying with hazardous waste, solid waste recycling, and transportation regulations often offsets any economic benefit companies might enjoy by trading materials and wastes. In the policy and regulations section of Industrial Ecology Opens New Opportunities for Government, we discuss these barriers and EPA's current and future goals for lifting some of these barriers when the Resource Conservation and Recovery Act (RCRA) is amended. As with off-site recycling, many of the public health and environmental risks are alleviated in an industrial park setting because of the close proximity of companies and information sharing among them.

A multidisciplinary team-based approach to process redesign allowed Fisher Scientific to identify 28 improvements that cost a total of $79,000, increased production yield an average of 28 percent, and saved the company a total of $529,000 a year.

(Romm 1994)

Benefits of P2

Companies that adopt P2 approaches are enjoying many benefits including reduced waste disposal and raw material costs, improved worker safety, and better public image.

Many firms have developed company-wide, systematic pollution prevention programs that include a detailed investigation and assessment process, employee training and incentives, and setting P2 goals.

Companies often adopt cleaner production techniques like switching from chlorinated solvents to less toxic nonchlorinated solvents with or without a P2 context.

In addition many states now mandate that companies submit detailed information about their waste minimization goals and progress.

Martin Marietta's Astronautics Group formulated a pollution prevention strategy. Their first project targeted TCA, the favored solvent for degreasing rocket components. By 1993 TCA vapor degreasing was completely eliminated. The up-front costs of the program included $70,000 in study and research plus $200,000 in implementation. The benefits include $50,000 yearly savings in material costs, $400,000 a year in avoided ozone depletion taxes through 1995, and $150,000 yearly reduction in operating, maintenance, and waste disposal costs.

(Romm 1994)

Challenges of P2

Some P2 practitioners see IE's goal of turning wastes into by-products as conflicting with P2's emphasis on source reduction, particularly of hazardous materials.

Present P2 based regulations sometimes make it impossible to reuse a waste, even between sites within a company.

Environmental costs are often hidden in company overhead, making it difficult to perceive the economic burden of business as usual or the savings from P2.

P2 Resources

Battelle P2 and LCA Institute for Pollution Prevention: http://www.estd.battelle.org/pp/.

DOE Pollution Prevention Information Clearinghouse, P2 Info Hotline: Telephone: 509-3P2-INFO or 509-372-4636, Internet: http://146.138.5.107/epic.htm.

EPA Technology Transfer, Cincinnati Research Lab, provides information, grants for pollution prevention, Telephone: 513-569-7562.

National Pollution Prevention Center for Higher Education: http://www.snre.umich.edu/nppc/.

Pacific Northwest Pollution Prevention Resource Center Environmental Home Page: http://pprc.pnl.gov/pprc/.

Pollution Prevention Information Exchange System (PIES) operates a hotline (703-821-4800).

Pollution Prevention Roundtable: http://es.inel.gov/aipp/.

Freeman, Harry, et al. 1992. "Industrial Pollution Prevention: A Critical Review." *Journal of American Air & Waste Management Association.*, June.

Romm, Joseph J. 1994. *Lean and Clean Management: How to Boost Profits and Productivity by Reducing Pollution.* Kodansha America, Inc., New York, pp. 131-3.

US-EPA Office of Pollution Prevention, 1991. *Progress on Reducing Industrial Pollutants.* October. EPA-21P-3003. US-EPA, Washington, DC 20460.

_______ Office of Research & Development. 1991. *Industrial Pollution Prevention Opportunities for the 1990s.* US-EPA, Washington, DC 20460.

_______ Office of Research & Development. 1992A. *Facility Pollution Prevention Guide.* May. EPA/600/R-92/088. US-EPA, Washington, DC 20460.

_______ Office of Research & Development. 1992. *Pollution Prevention Case Studies Compendium.* US-EPA, Washington, DC 20460.

_______ Waste Minimization Branch. 1994. *Review of Industrial Waste Exchanges.* Office of Solid Waste. EPA-530-K-94-003.

DESIGN OF ORGANIZATIONS

An organization adopting industrial ecology as a framework for its operations can benefit from exploring its implications for all elements of its management: mission, culture, organizational structure, incentives, financial and management accounting, employee education and training, supplier and customer relations, and information systems.

The possible benefits are increased competitiveness (or effectiveness, in the case of public organizations), greater capability to put IE's systems approach into action, and deeper insight into opportunities for major reductions in environmental impact.

Fortunately there are strong parallels between the sort of organization needed to implement IE and present trends in organizational design. The learning organization, socio-technical systems work design, systems dynamics, the more robust versions of total quality management, and the viable system model are a few of the design methods that have a natural affinity with industrial ecology. From different angles, each method seeks to build an organization that is

- Creative in forming a compelling vision and continuing to evolve it
- Adaptive, resilient, and proactive

- Able to view itself as a dynamic, living system operating in its business and natural environments
- Skilled at managing complex flows of information to discern possible surprises as well as likely trends
- Competent at balancing strategies to meet present requirements and future needs or conditions
- Able to mobilize the skills and knowledge of employees at all levels of the organization.

Aligned with industrial ecology, these organization design methods could help channel change into awareness of the organization as a direct participant in natural systems. Internally, they could support the integration of many specific initiatives into a functioning whole system. Some of these organizational initiatives include:

- Evolve the organizational mission and broad strategy to support achievement of IE objectives.
- Set environmental performance objectives at each level of organization.
- Adapt financial and management accounting systems to reflect environmental costs of operations.
- Change incentive systems to reward innovations that help achieve IE objectives.
- Balance R&D spending to assure new products and processes for achieving IE objectives.
- Provide employee education and training in systems thinking, industrial ecology, and technical and organizational innovation.
- Offer environmental leadership and education to suppliers and customers.
- Use information systems to provide feedback on environmental performance relative to organizational objectives, local and global environmental conditions, and innovative solutions.

Organizations in policy, finance, education, and other aspects of "soft" infrastructure need similar redesign in order to better support and guide manufacturing and service companies, hard infrastructure firms and agencies, and consumers. Consultants in strategic environmental management (SEM) emphasize many options like these,

S.C. Johnson, a producer of household cleaning products, developed a series of green packaging initiatives in 1990. These included an "enviro-box," reducing materials 92 percent; increased recyclable content in packaging; and development of recyclable bottles for some products.

To further enhance environmental performance, the company communicated its long-term goals to all suppliers and invited them to attend a high-level conference. Heads of Johnson's three major business units attended to reinforce the importance of the mission. Representatives of some of the company's key customers also attended to report their environmental expectations and to build Johnson's "message of increasing environmental interdependence among companies along the entire product chain."

The follow-up to this event has seen acceleration of the process of introducing further packaging improvements and the elimination of specific chemicals with adverse environmental effects.

(Schmidheiny 1992)

but often do not bring an adequate systems view to the task of integrating organization-wide change.

One might ask if the business schools and management science aren't already covering the field of organization design we have outlined. Tom Gladwin, a central proponent of sustainable development in business management, says:

> . . . most management theorizing and research continues to proceed as if organizations lack biophysical foundations. Organic and biotic limits in the natural world are excluded from the realm of organizational science. . . . Phrases such as biosphere, environmental quality, ecosystem or sustainable development are virtually absent from the leading management journals (Gladwin 1995).

Relatively few business schools have even elementary courses on environmental management. In many corporations, management practice has moved far ahead of the academy in considering environmental issues as a core strategic concern.

Information Systems

A fundamental component of organizational design will be effective information and communications systems to support IE work within and between organizations.

Some requirements and capabilities needed in such systems include:

- Clear perception of complex fields of information and data, including indications of level of reliability of information
- Easy shifting of time and space dimensions in viewing information
- Accessibility of information to generalists or laypersons as well as specialists
- Ability to test alternative strategies through what-if analysis
- Clear statement of values and assumptions implicit in information and data
- Ability of information systems to "learn from experience"
- Timely feedback on performance.

New Organizational Relationships

In many ways, industrial ecologists assume a collaborative model of business. There is a great deal to support this assumption in recent trends for both large and small firms. Many companies, including direct competitors, are recognizing strategic partnering as a key source of competitive advantage. For instance, Canon co-invested with Hewlett-Packard (HP) in advancing laser printer technology in the 1970s. The company now supplies HP with laser print engines but continues to compete in inkjet printers.

Large corporations often ally closely with smaller companies to ensure a critical supply or even to conduct research and development:

- GM has struck deals with electric car manufacturers to gain the flexibility and speed of innovation an entrepreneurial venture affords.
- In 1991 S.C. Johnson, a producer of household cleaning materials, brought environmental excellence into its Partners in Quality program with 70 top suppliers.
- To prepare for European legislation requiring auto manufacturers to take back their products, Renault and BMW formed a partnership to research disassembly and recycling technology and to share networks of recycling companies.

An advertisement from Nippon Steel, a major Japanese corporation, expresses this spirit of corporate partnership in explicitly ecological language (using the same term companies at Kalundborg, Denmark, have chosen to describe their relationship):

"Symbiosis is one of the natural world's truly beautiful systems. In reality, this principle of dynamic natural relationships exists not only among plants and animals, it also applies to animals and humans, humans and humans, companies and companies, companies and the environment, humans and the earth. It is this very relationship, expressed in the term 'symbiosis,' that is our goal for bringing about better business global partnerships. Through both free competition and harmony, based on a spirit of mutual benefit and trust, Nippon Steel will continue to make these interactions more productive and fruitful for our lives."

Small to midsize firms are forming alliances variously called value-adding partnerships, flexible networks, or manufacturing networks. This form of partnership relies on a hub company that coordinates marketing and in some cases research and purchasing for the loosely bound network.

A growing number of companies are contracting to purchase material and energy by-products in one-to-one partnerships. For example, US Gypsum buys gypsum from utilities (where it is generated in flue gas desulfurization scrubbers) and paper from recycled paper companies for its completely recycled sheet rock.

Resources on Organizational Design

Ackoff, R.L. 1981. *Creating the Corporate Future*. John Wiley & Sons, New York.

Ackoff, R.L. 1984. *A Guide for Controlling Your Corporation's Future*. John Wiley & Sons, New York.

Argyris, Chris, and Schon, Donald A. 1978. *Organizational Learning: A Theory of Action Perspective*. Addison Wesley Publishing, Reading, Massachusetts.

Beer, Stafford. 1979. *The Heart of Enterprise.* New York, NY: John Wiley & Sons. (Beer's Viable System Model offers a dynamic organizational structure grounded in the understanding that organizations are living systems interacting with larger living systems. It is a vital tool for managing the transition to industrial ecology.)

Beer, Stafford. 1985. *Diagnosing the System for Organizations.* John Wiley & Sons, New York, NY.

Business and the Environment. 1992. "Renault and BMW Sign Recycling Coop Agreement," October, p. 8.

Clemson, Barry, and Lowe, Ernest. 1993. "Total Quality Management and Comprehensive Change." *Engineering Management Journal.* Vol. 5, No. 1 (Best Paper Award at 1992 ASEM-IEEE Conference).

Clemson, Barry, and Lowe, Ernest. 1993. "Which Way to Rome: Choosing a Path for Change." Engineering *Management Journal.* Vol. 5, No. 4, December.

Denison, Daniel R. 1990. *Corporate Culture and Organizational Effectiveness*. John Wiley & Sons, New York.

Emery, Fred, and Trist, Eric. 1973. *Towards a Social Ecology*. London, Plenum Books. (On sociotechnical systems design of workflows.)

Gladwin, Thomas, et al. 1995. "Shifting Paradigms for Sustainable Development: Implications for Management Theory and Research." *The Academy of Management Review*. Vol. 20, No. 4, October.

Hatch, Richard. 1991. "The Power of Manufacturing Networks." *Transatlantic Perspectives*. Number 22, Winter,. pp. 3-6.

McWhinney, Will. 1992. *Paths of Change: Strategic Choices for Organizations and Society*. Sage Publications.

Pasmore, William, and Sherwood, John (eds.). 1978. *Sociotechnical Systems: A Sourcebook*. University Associates, San Diego.

Pinchot, Gifford, and Pinchot, Elizabeth. 1993. *The End of Bureaucracy and the Rise of the Intelligent Organization*. Berrett-Koehler Publishers, San Francisco.

Schmidheiny, Stephan. 1992. *Changing Course*. The MIT Press, Cambridge, MA, pp. 217-220.

Senge, Peter. 1990. *The Fifth Discipline: The Art and Practice of the Learning Organization*. Doubleday/Currency, New York.

Senge, Peter. 1994. *The Fifth Discipline Fieldbook: Strategies and Tools for Building a Learning Organization.* Doubleday/Currency, New York.

Weisbord, Marvin, 1987. *Productive Workplace: Organizing and Managing for Dignity, Meaning, and Community*. Jossey-Bass, San Francisco.

THE DIALOGUE BETWEEN IE AS RESEARCH AND IE AS CONTEXT FOR ACTION

From a research perspective, IE is defined as follows: "Industrial ecology is the study of the flows of materials and energy in industrial and consumer activities, of the effects of these flows on the environment, and of the influences of economic, political, regulatory, and social factors on the flow, use, and transformation of resources" (Robert M. White, President, National Academy of Engineering).

Some major aspects of this research include:

- Integrated examination of environmental threats such as persistence of toxic chemicals, synergistic interactions among toxics, depletion of resources, and disruption of planetary life-supporting cycles.
- Better understanding of the metabolism (use and transformation) of materials and energy in industrial ecosystems. Better information about potential waste sources and uses.
- Improved social and economic mechanisms that encourage systems optimization of materials and energy use (markets, policies, incentives, and regulatory structures).
- Consideration of the role of households and communities in generation of waste and pollution and in creating a closed-loop industrial system.

Business and governmental organizations are now beginning to gain value from working in this new context. Their experience in applying IE will provide practical experience to guide research. Without it, academic and government research in this field will be barren. So we need a continuing dialogue between researchers and industrial ecologists applying its methods and tools in corporations and governments at every level.

See Chapter 6 for further discussion of IE research under Research and Technology Policy and Chapter 8, Link Action and Research.

LEVELS OF APPLYING IE IN TRANSPORTATION: THREE CASES

Incremental improvement in the design of automobiles, proposed radical redesign of small vehicles (the Hypercar concept), and plans for an integrated railbased transportation system illustrates the broad spectrum of design opportunities for industrial ecology. Each case is at a different system level and in a different time frame.

Automobile Re-Design Within the Current Product Concept

An industrial ecologist would use tools such as design for environment to support short-term enhancements in automobile design. The basic question would be: How can we optimize trade-offs to reduce energy use and pollution in the production process as well as in use of the product?

The automobile and its infrastructure, as now designed, is a major environmental threat. The production of cars and their fuel pollute land, sea, and air. Autos are a major contributor to atmospheric ozone pollution, generate 18 percent of global carbon dioxide emissions, and are a significant source of water pollution. Autos use nonrenewable petroleum at an ever increasing rate. Roads, freeways, and parking garages typically use between one-third and one-half of urban space and a major share of transportation dollars. Interstate highways cover valuable agricultural land, while air pollution damages crops and forests.

The automobile industry has made serious efforts to prevent pollution in manufacturing processes, decrease materials wastes, and practice internal and external recycling of metals and plastics (process design). Product design seeks improved fuel efficiencies and lower emissions, usually through redesign of the internal combustion engine and specific innovations such as the catalytic converter, electronic fuel injection, and composite materials.

Designers are also seeking to greatly enhance recyclability of automobile components. Anticipated European take-back legislation is prompting automakers there to begin design for disassembly and to team together in setting up an infrastructure for realizing value from cars returned to them (Klimisch 1994).

Change in design in the U.S. is largely driven by Clean Air Act revisions, CAFE mileage standards, zero emissions vehicle legislation in California and the Northeast, and Germany's proposed take-back legislation (placing responsibility on manufacturers for recycling at the end of a car's life). But the targets are still generally modest, given the magnitude of the automobile's burden on the environment. The redesign is incremental change in the basic familiar product. Even Big Three electric car designs to meet California's 1998 zero-emissions deadline are retrofittings of existing internal combustion models (The mandated standards call for 2 percent of fleet sales in the state to be zero-emissions vehicles.)

This incremental level of DFE is necessary to develop improved products in a time of transition. Business constraints on more fundamental innovation include the large investment in traditional production facilities, lack of strong demand for change from markets (in North America as well as those opening in Asia), and relatively modest performance standards set in public policy. Continuing to apply DFE at this level may give very wise decisions about very wrong long-term choices. Most solutions are likely to be short-lived.

The RMI Hypercar

At another level, an industrial ecologist might ask, how can we transform small vehicle design to capture levels of efficiency and freedom from pollution not possible within the existing internal combustion model.

Amory and Hunter Lovins have challenged the automobile industry to stretch to another level of design with the Rocky Mountain Institute (RMI) Hypercar proposal. The Lovins challenged a team at RMI to go to the basic physics and engineering of small vehicle design. Forget the common wisdom of the industry. The result is a design with radical implications for fuel and materials efficiency, emissions, manufacturing process, and the nature of the business itself.

The Hypercar is projected as an ultralight, highly aerodynamic vehicle powered by a small electricity generating engine (gasoline, liquid gas, or hydrogen cell). The engine transmits current to drive mechanisms in the wheels, which also recapture energy from braking (~70 percent of braking energy). Highly energy efficient accessories (lights, heating, A/C, radio) also reduce energy demand. Body and frame design with strong composite materials provide passenger safety higher than in traditional vehicles at almost one-third the weight. Selective use of superstrong carbon fibre and other composites will reduce the number of body parts and simplify production and assembly.

The RMI performance models compute that Hypercars will be one hundred times cleaner than present cars (or pure electrics) and operate at 150 mpg in the near term design. (GM's Ultralite concept car has demonstrated only 62 mpg.) Potential fuel efficiencies could double and triple that high level with more advanced designs. The total energy draw will equal that used for accessories in present autos.

RMI indicates that the Hypercar design concept will enable an equally radical transformation of automobile production. "Moulded composite cars need much less and vastly cheaper tooling. The tooling's short life and very quick fabrication supports fast cycles, short time-to-market, continuous improvement, small production runs, and strong product differentiation—a striking strategic advantage." Conceivably, the U.S. Big Three and other global auto companies could be left in the dust of entrepreneurs who see the parallels with the emergence of a personal computer industry (coming out of garages!), a new industry able to rapidly challenge the mainframe manufacturers in the early 80s.

A more likely scenario is that advanced design concepts such as the RMI Hypercar will be developed and tested through a partnership between startups and major automakers. Already many entrepreneurial companies are growing in the market niche created by California's zero-emissions standards, and some are collaborating with the major producers.

The Lovins claim that the technological foundation for realizing the Hypercar concept has been laid. The design challenge is in breaking out of the conventional wisdom of how we conceive and build cars and in integrating the breakthroughs already made.

With the Hypercar we move from incremental to transformative change, the realm of industrial ecology. At this level we start designing on a clean slate, asking what do we really need to do to provide the customer utility we're here for. We are open to fundamental redesign of product, production processes, and the very nature of our business.

An Integrated Transportation System

At a still broader level, an industrial ecologist would ask, how can we design integrated transportation systems to move people with highest resource efficiency and lowest possible pollution? How can telecommunications, urban planning, and work design reduce the number of trips and distance traveled.

Amory Lovins also recognizes that even transformative design at the level of individual vehicles gives no more than a partial, necessary solution. A necessary and sufficient solution will only be found at a higher level of design, one that addresses the system in which automobiles function. Hypercars and smart highways alone will still leave us with an overall system design burdening the environment via materials and land use. One need only look at the implications of deploying an automobile infrastructure in China to grasp the dimensions of this issue. Agricultural land there is already being paved over for industrial development. It is degrading heavily from desertification, pollution, and reliance on chemical fertilizers.

Suntrain, an entrepreneurial start-up company in California, has moved design for the environment to the level of the system needed for an adequate response to the environmental challenge: intermodal, railbased transportation. Christopher Swan has designed the technical, business, financial, and political infrastructure for transforming urban and interurban transportation.

Imagine being able to make one phone call to route, schedule, and pay for a trip via public transit that would get you from here to there faster than in your car. Perhaps you start via a van that picks you up at your door, connect without waiting to a subway, transfer to an electric or gas powered rail car for the main trip, and pick up a rental Hypercar to make business calls at the other end. No gridlock. Time to read, tap into the Internet, or work on that unfinished spread sheet.

Total cost to the individual for using this integrated system for business and recreational uses: 50 percent of the cost for operating and maintaining a personal vehicle to cover those miles.

Environmental benefits: An integrated transportation system such as the Suntrain project would dramatically reduce fuel use, emissions, and resource use. One railway passenger car worth two million dollars lasts 20 years, and replaces the mileage consumed by six thousand automobiles worth ninety million. Less urban and rural land would be consumed by the system and pavement could be removed in some areas.

Is this pie-in-the-sky fantasy? Even without the one call information system and tight connections, passenger rail transport has increased significantly on selected U.S. routes in the last decade. As with the Hypercar, the technologies are fully available now, including self-propelled passenger cars, power plants using alternative fuels, and distributed information systems for system management and customer travel efficiency. Geopositioning satellites are now used by railways in tighter, more efficient scheduling of trains. Swan says 250,000 miles of rail track are used at roughly 15 percent of capacity (a sunk infrastructure worth one trillion dollars).

With Suntrain we move to the design of institutions and infrastructure grounded in the customers' need to move effectively in the short-haul ranges (up to 400 miles) that constitute the majority of passenger trips. Rather than focus on a specific mode of transportation, the concept addresses the need for an integrated system, linking all modes. The concept also unifies meeting the customer's business, commuting, recreational, and person needs.

The Suntrain vision reflects a higher level of design for environment: design at the level of business concept and social system that integrates already present technological innovations into a new solution.

Critics often raise the question, "How will we ever get people out of their cars?" If the built-in incentives don't achieve this, then we move to a level of design vital to the transformation in personal lifestyles sustainability demands. This is creation of the social incentives, communications channels, and means of learning through which people can freely adopt new patterns of behavior.

This case offers a good illustration of the service economy concepts of Walter Stahel discussed in the next chapter. Historically, the railroad industry has demonstrated many product-life extension strategies he proposes.

Resources on Transportation

Rocky Mountain Institute, 1739 Snowmass Creek Rd., Old Snowmass, CO 81654-9199. Telephone: 303-927-3851, Fax: 303-927-4178.

Klimisch, Richard L. 1994. "Designing the Modem Automobile for Recycling." In Allenby, Braden R., and Richards, Deanna (eds.). *The Greening of Industrial Ecosystems*. National Academy Press. Washington, DC.

Lovins, Amory B., Barnett, John W., and Lovins, L. Hunter. 1993. *Supercars: The Coming Light-Vehicle Revolution.* Rocky Mountain Institute, Snowmass, CO.

Lovins, Amory B., and Lovins, L. Hunter. 1995. "Reinventing the Wheels." *Atlantic Monthly*. January.

Renner, Michael. 1988. *Rethinking the Role of the Automobile*. Worldwatch Institute, Worldwatch Paper 84, Washington, DC.

Swan, Christopher. 1995. Suntrain Inc. Business Plan. San Francisco.

Swan, Christopher. 1996. *Transportation Transformation.* Ten Speed Press, Berkeley [In press].

4 Product-Life Extension and the Service Economy

INTRODUCTION

An advanced strategy for lowering pollution and the demand for energy and materials is designing durable and upgradable products with a long life span. The Swiss researcher, Walter Stahel, has pioneered this strategy, which poses a possible Catch 22: How can manufacturing companies remain profitable if their products last so long? Stahel suggests that they refocus their mission to delivering customer service (selling results, performance, and satisfaction, rather than products) and owning the equipment themselves as the means of providing this service.

The concepts of product-life extension and the service economy go beyond all other IE approaches to closing the loop in industrial/consumer systems. They are an essential complement to the work of industrial metabolism, design for the environment, and other IE methods.

Is Stahel's vision useful only in the long-term? Several major companies are now moving in this direction. Stahel's concepts could possibly be used by major entrepreneurs to enter markets now dominated by existing companies.

Xerox Asset Recycle Management Program

The ARM initiative at Xerox Corporation reflects the equipment design strategies recommended by Walter Stahel. The mission: "Asset Recycle Management provides the leadership, strategy, design principles, operational and technical support to maximize global recycling of parts and equipment, resulting in a major competitive, as well as environmental advantage for Xerox."

This mission has been built into the company's global organization with an ARM Vice President responsible for achieving 100 percent recyclability of all manufactured parts and assemblies. Remanufacturing to high quality standards and resale to new users will extend the life of equipment several fold and reduce demand on virgin resources. "Xerox chooses to evaluate products on the basis of quality and performance, not on the degree of virgin material used . . ." The initiative is designed to streamline the process by which returned machines are reconditioned, thus increasing return on investment.

The company estimates it has added hundreds of millions of dollars to its bottom line since ARM was formally started in 1991.

(Xerox 1995)

PRODUCT-LIFE EXTENSION CONCEPTS

Walter Stahel has linked the concerns of industrial metabolism and DFE with a broader level of design: the basic mission of a business. Stahel, a director of the Swiss Product-Life Institute, argues that closing loops through recycling is only a partial solution. It does not slow the rapid and unsustainable flow of materials and goods through economies.

He proposes product-life extension as the necessary complement to recycling, and he suggests business strategies for achieving it and sketches the dimensions of a service oriented economy. While his vision implies significant changes, Stahel identifies major corporations such as Xerox, Schindler (the world's second largest elevator company), Agfa Gevaert, and Siemens that now demonstrate the concept in practice. *See sidebars on Xerox on this page and on Agfa Gevaert in the business opportunities section of the Executive Briefing.*

Product-life extension implies a fundamental shift from selling products themselves to selling the utilization of products, the customer value they yield. This change in the source of economic value to firms depends upon enhancing product life through several key design strategies. Designers would seek to optimize products for the following qualities:

- Products that are durable and difficult to damage, modular, and multifunctional[1]
- Products and subcomponents that are standardized, self-repairing, and easy to repair or upgrade
- Components that can be reused in new systems
- Units or systems that can be easily reconditioned and remanufactured
- A distributed service network that is developed for easy maintenance and upgrading of systems.

These design strategies are already part of the design for the environment toolkit. They would significantly help achieve the central industrial metabolism objectives of

[1]Several manufacturers are now marketing products integrating a laser printer, scanner, fax, and telephone in one device.

cutting demand on material and energy resources and reducing pollution from manufacturing.

The Service Economy

As a company moved from maximizing sales of material products to the delivery of customer satisfaction, its long-term source of competitive advantage would become the ability to provide the needed service. Revenues could come from leasing equipment with a long life; continuing maintenance and service; major upgrading of systems, parts, and supplies; and service provider training and licensing.[2] Or the company might simplify the transaction by offering one use-based fee.

Stahel argues that *if the company is compensated on the basis of service provided, then employees will have strong incentives to minimize materials and energy used in the systems that deliver the service to the customer.*

Stahel also considers the larger transition to a decentralized and skill-based service economy that product durability implies. Economic value would be based on utilization (customer satisfaction in the service gained) rather than exchange. Decentralized labor-intensive service centers would create many skilled jobs for workers no longer needed in centralized, automated production units. Resource use would be lowered as products no longer moved rapidly from factory to customer to landfill.

Walter Stahel's work represents design at the level where a company asks, *"What business are we really in?"* Decisions at this level may have the greatest impact on a firm's environmental performance. Stahel's concepts could guide existing corporations in redefining their mission. In some industries, entrepreneurs could adopt this systems approach in order to compete with established firms. We described one proposed example of an entrepreneurial transportation system reflecting Stahel's thinking in the case at the end of the previous chapter. At the end of this chapter is a future scenario describing a home services firm. The future scenario of a country applying industrial

[2]Service might be provided through a decentralized service network, licensed and trained by the central company. Economically, this would be an important source of skilled jobs.

ecology at the end of the book illustrates applications in telecommunications and transportation.

Benefits of Product-Life Extension

Product-life extension is a strategy for developing very large reductions in materials and energy use needed to satisfy growing consumer needs within limits on resources. Stahel estimates that it could increase the productivity per unit of resource used ten fold. The strategy includes strong economic incentives for achieving these objectives. This improved resource productivity translates to increased profitability and competitiveness.

A service economy offers a decentralized means of developing skilled jobs. Stahel's systems approach could give independent entrepreneurial ventures competitive advantage in entering markets when major corporations who decline to use it remain focused on selling products.

Challenges of Product-Life Expansion

This approach to sustainability requires long-range vision and major organizational and technological redesign on the part of corporations. (Investment markets' present focus on short-term financial performance does not support such fundamental change.)

Companies adopting product-life extension run the risk of making major investments in technologies for service delivery that may become outmoded. To what extent can modular design for easy upgrading offset this risk?

RESOURCES ON PRODUCT-LIFE EXTENSION

Walter Stahel, Product-Life Institute, 18-20 Chemin Rieu, CH-1208 Genève, Switzerland. Telephone: (022) 346-39-32, Fax: 347-20-78.

UNEP Working Group on Sustainable Product Design, 3rd Floor, Building B, Nieuwe Achtergracht 166, 1018 WV Amsterdam, The Netherlands. Telephone: (31) 20 525 6268, Fax: 625 8843. E-mail: unep@unep.frw.uva.nl. Internet-http:// unep.frw.uva.nl. This UNEP project considers the service and function that a product provides rather than just maximizing environmental concerns related to existing products. "Often the whole system in which a product exists and contributes is more important to achieving sustainability than the product itself." The project includes a research program and an international network of researchers, consultants, and designers.

Borlin, M. 1990. "Swiss Case-Studies of Product Durability Strategy." Product-Life Institute. Genève, Switzerland.

Giarini, Orio, and Stahel, Walter. 1989/1993. *The Limits to Certainty: Facing Risks in the New Service Economy*. Kluwer Academic Publishers, Dordrecht and Boston, MA.

Stahel, Walter. 1986. "Product-Life as a Variable: The Notion of Utilization." *Science and Public Policy*.13(4), August.

Stahel, Walter. 1994. "The Utilization-Focused Service Economy: Resource Efficiency and Product-Life Extension." In Allenby and Richards, *The Greening of Industrial Ecosystems*. National Academy of Engineering, Washington, DC. Available through the National Academy Press Office (202-334-3313).

Stahel, Walter R. 1995. "The Functional Economy and Cultural and Organizational Change." In Richards, Deanna J., *The Industrial Green Game: Implications for Environmental Design and Management*. National Academy Press, Washington, DC. Available through the National Academy Press Office (202-334-3313 or 1-800-642-6242).

Stahel, Walter, and Reday, G. 1981. *Jobs for Tomorrow, the Potential for Substituting Manpower for Energy*. Commission of the European Community. Vantage Press, NY.

Victory, Kathleen. 1995. "Focus Report, Why Smart Companies Will Take Part in the Debate on Sustainable Production and Consumption." *Business and The Environment.* Vol. 6, No. 8. Arlington, MA, August.

Xerox. 1995. *CSS/ISC Asset Recycle Management*. June. Revision 2.0. For information on this program, write Asset Recycle Management, Xerox Corporation, 455 West Commercial St., East Rochester, NY 14445.

A FUTURE SCENARIO: Gambit Ltd. Markets Integrated Household Service Systems

October 10, 2003: Gambit Ltd. celebrates its fourth year of successful growth today. In honor of the occasion, Sarah Hemmings, CEO, said, "We created a niche and so far we're filling 90 percent of it." The firm was recently acquired by General House, the major appliance manufacturer who joint ventured in Gambit's startup.

The rapidly growing firm provides new homeowners with integrated service systems that fulfill all lighting, heating, cooling, ventilating, and kitchen needs. Gambit owns the hardware, installed by local contractors that it licenses as Gambit service centers. The householder pays a monthly lease fee based upon usage. The fee covers use and service of the system and energy costs.

By creating a whole system to meet all lighting, heating, and cooling needs, the company's engineers have cut the Gambit customer's total energy draw by 65 percent. All components are highly energy efficient. Traditionally wasted heat from any one function is reused for other heating or cooling purposes, depending upon the season. The system's control center includes occupancy sensors that trigger dimming of lights when a room is empty.

Gambit offers technical services to real estate developers needing support in building with appropriate insulation, use of photovoltaic building components, daylighting, and other energy efficient design options. The design of both the home and the Gambit system allow for upgrades as new technologies become available.

The first residential developers to risk working with Gambit's radical approach were a mix of firms committed to sustainable design and others who simply saw a new source of competitive advantage. By cutting their major cost of heating and cooling equipment, developers were able to offer houses and condos at a highly competitive price.

First-home buyers needed a clear introduction to the savings they could gain by the shift from owning to leasing. Gambit's initial demonstration projects had created a track record to support sales as well as the shakedown of the technologies involved.

Within two years, Gambil International had started installing systems in developing country public housing projects. General House CEO, Martin Agruerro, initiated the joint venture with Gambit in 1997 as a calculated risk. If the new company could break into the market, GH would be manufacturing its equipment. If it were successful, Agruerro saw that Gambit might be creating his corporation's future. One recent move by GH indicates this may be happening. The firm has also acquired Home Telecom Systems, the breakthrough venture that has paralleled Gambit's strategy by delivering integrated communications, information, and entertainment services.

5 Emerging Business Opportunities Suggested by Industrial Ecology

INTRODUCTION

Industrial ecology provides powerful opportunities for business, including cost savings, new product lines, and new customers. This potential will grow as the approach becomes commonly accepted. Some major realms of IE opportunity include:

- Introducing cost savings and new revenues in existing operations
- Entering new markets for existing goods and services
- Marketing emerging technologies, materials, and processes
- Supporting the organizational change, technical, and information needs of an IE-based economy
- Integrating technologies and methods into innovative new systems.

We will discuss a sampling of opportunities in each of these areas.

COST SAVINGS AND NEW REVENUES IN EXISTING OPERATIONS

Environment and business newsletters and magazines are full of stories of cost savings from pollution prevention

"We decided to take a much more holistic view of our procedures, products, and services," said Steelcase Inc. manager of corporate environmental quality, David Rinard. This Michigan furniture manufacturer is seeking to "build industrial ecology into the design of a product from the start, rather than deal with what comes out at the end of the pipeline." Company managers consider the environmental implications of every business decision.

Steelcase has embodied this perspective in several initiatives. It has created a subsidiary, Revest, to refurbish used office furniture and sell it in secondary markets. A "redirected work center" employs partially disabled workers to create products out of recycled materials from primary production lines. The company also found that shipping products without packaging actually resulted in less damage. "If workers see they are handling high-quality furniture they tend to be more careful."

(Bureau of National Affairs 1994. Daily Report for Executives, April 14, 1994. Online version)

programs. These are generally achieved through innovations within a plant or between units of a single business. Industrial ecology extends the borders for this opportunity with savings and revenues possible from interfirm collaboration. IE also provides ways of seeing new opportunities within a company or its individual plants.

IE champions have tended to emphasize one particular source of advantage: turning energy, water, and material wastes into by-products. Common sense, not any awareness of IE has prompted hundreds of firms to negotiate contracts for trading former wastes.

For instance, utilities sell gypsum from scrubbers to wallboard plants and fly-ash to cement companies. Their bottom line shows new revenues and savings from reduced waste transportation and landfill costs. Sixty percent of the heat generated in a typical power plant goes up the stacks. As steam or hot water this can be used for industrial or agricultural processes. In Kalundborg, Denmark, a series of bilateral deals for exchanges between plants has earned $120 million on an investment of $60 million in infrastructure. *See Chapter 7 on Industrial Ecosystems for more detail on the energy, water, and materials flows at Kalundborg.*

Douglas Holmes, a Massachusetts chemical engineer, has proposed that companies should treat every "waste" as a product with a name, stock number, and a production and advertising budget. *Then financial reports would clearly show the costs of manufacturing products a company hasn't yet learned how to sell.* Motivation would increase to either stop producing them or to discover new markets for them.

Perceiving wastes as by-products is only one opportunity for cost-savings that industrial ecology offers. IE's systems view and methods give managers and designers a means of assessing competing solutions to environmental problems, weighing both environmental and economic issues. With this assessment they can better define the most effective investment to make.

For instance, an industrial metabolism study of the energy, materials, and water that flows from suppliers, through a plant, and out to customers and the environment would highlight areas where investments could create the greatest gains in efficiency. In some cases, these might actually be outside the plant, in supplier firms, or customer patterns of use.

FINDING NEW MARKETS FOR EXISTING GOODS AND SERVICES

Many firms will discover new markets for their existing products in companies, communities, and agencies applying industrial ecology. We will offer examples in two areas: opportunities in by-product exchange and in restoration and management of landscaping.

By-product trading among companies opens a number of new business niches for reprocessing and information technologies and for service providers. Producers of a wide range of materials processing equipment (for grinding, sifting, sorting, purifying, and packaging), for instance, will find new markets. Intermediary companies altering wastes mechanically or chemically to meet customer specifications will need this equipment as the concept of waste is eliminated. Another business finding new markets will be manufacturers of equipment to channel flows of materials, water, or heat among plants. Brokerage or investment recovery firms may find a profitable role in optimizing by-product exchanges in a network of plants. They will need information systems supporting this optimization.

The "Decomposer" Niche: Raymond Côté at Dalhousie University has identified the need for potential "decomposer" companies in the industrial park at Burnside. (These companies are metaphoric equivalents of the organisms in an ecosystem that turn dead animal and vegetable materials into a consumable form.) The decomposer niche includes companies that consume otherwise unusable wastes, processing them into usable feedstocks, and companies that take apart equipment and market reusable components and materials.

In terms of technologies, decomposer plants would utilize recovery technologies that allow extraction of valuable materials from waste streams (e.g., extracting metals from liquid or solid wastes); and recycling technologies that prepare a by-product for reuse (e.g., shredding plastics). These processor companies play a useful intermediate role and are especially important where they accumulate small flows of residues from companies and generate economically useful masses. A solvent recycler is one example.

Investment recovery: A new field of management, investment recovery, takes a systemic approach to ending waste. It is "an integrated business process that identifies,

The SOB who welded the floor drains shut.

In the early 1990s, Hyde Tools in Massachusetts set a target of zero water discharge within five years. By mid-1993 they had cut 27 million gallons of waterflow to 1.2 million gallons. Getting there involved having Doug DeVries, environmental and training manager, become known in the plant as "the SOB who welded the floor drains shut."

DeVries said, ". . . we had developed a process that didn't require any (water) discharge. But . . . every Saturday they would dump a tank of wastewater. We tried the gentle approach, which people remembered for about a week and a half. Then, we decided that if they had no place to put the water, they'd have to change their actions. So, we welded the floor drains shut. At the approach of the weekend, people recognized that they had to turn on the filtration unit. It's a matter of breaking people's habits."

Hyde replaced clay absorbents with ground corn cobs to pick up oil and water. They use shipping pallets made from molded wood waste, that burn with zero ash. They buy recycled materials, such as 100 percent post-consumer paper "peanuts" dunnage,and save a tip fee of $100 per ton. They went from sending 176 tons of solid waste to the landfill down to 35 tons. By the end of 1993 they expect to have zero input to landfills in Massachusetts. Since the company started its pollution prevention efforts in early 1989, it has invested about $100,000 and saved approximately $250.000.

(*Business and the Environment*, August 1993)

for redeployment, recycling, or remarketing, nonproductive assets generated in the normal course of business" (Phillips Petroleum, undated). Assets include:

- Idle, obsolete, unused, or inoperable equipment, machinery, and facilities
- Excess raw materials, operating inventories, and supplies
- Construction project residues
- Machinery, equipment, and fixtures in facilities scheduled for demolition
- Offgrade, out-of-specification or discontinued products
- Waste stream and process by-products.

Investment recovery develops strategies and procedures to recapture the highest value from all surplus assets. It seeks ways to reduce operating and disposal costs, ways of preventing the waste, and markets for redistributing the by-products for increased economic value. Practitioners work with a broad range of projects, from industrial plant or military base closings to disasters. Phillips Petroleum, BASF-US, and MCI Telecommunications are among the companies that have adopted this approach.

An investment recovery firm could integrate the above functions into a comprehensive strategy for an industrial park, for an industrial area, or for a whole community. If the firm's fees are based on a retainer plus percent of savings and/or revenues it would have an incentive for supporting this holistic approach.

An IR firm could also negotiate technical assistance from universities, federal labs, and environmental pollution prevention and energy efficiency programs. With client companies it would target unutilized supplies that need this research input in order to be marketable. The firm could also recruit the right providers and take on the role of negotiating with regulatory agencies to define conditions for reuse of materials presently restricted.

Product remanufacturing: Remanufacturing firms serve an environmental objective by extending the life of products (and the energy and materials invested in them). If the rehabilitation process includes upgrades for energy efficiency, process control, and pollution prevention, the resulting products demonstrate further environmental and economic benefits. Remanufacture of capital equipment (for manufacturing, transportation, medical uses, or other sectors) offers potentially strong export opportunities. Many newly developing countries (such as Vietnam) seek to leverage limited capital by purchasing used capital equipment.

Ecosystem restoration: Companies who do restoration of ecosystems are finding opportunities with corporations and public infrastructure agencies. A familiar instance is mitigation of development impacts on one ecosystem by restoring a "comparable" ecosystem in another location.

As organizations adopt an IE perspective, they are likely to see further value in restoration projects to supplant traditional landscaping of company lands. This alternative cuts costs, can modify climatic impacts, and earns a genuine green image. The restoration furthers the ecological value of recreating habitats and opens educational experiences for employees and residents of the area.

MARKETING EMERGING TECHNOLOGIES, MATERIALS, AND PROCESSES

Full application of industrial ecology will also require new technologies, materials, and processes. Firms in the market niches described in the previous section would also be drawing upon new products. Reprocessing of by-products may require new chemical or mechanical processes. Product disassembly will require advanced robotics. Commercialization of biopolymers will supplant nonrenewable sources in many materials applications.

Technology Evaluation

Industrial ecology provides means for evaluating technical innovations, for understanding how they can most effectively be deployed, and possibly for creating cost-effective uses earlier in the development cycle. This suggests business opportunities in evaluation and systems design for environmental, industrial metabolism, etc.

Public works managers, for instance, could use support in assessing the claims of competing salesmen offering solid waste management systems. At a higher level, they might benefit from examining broader strategies for

Interface Inc.—Sustainable Floor Coverings

With worldwide sales of $1 billion, Interface Inc. of Atlanta, Georgia, is a recognized leader in the worldwide commercial interiors market, offering floorcoverings, fabrics, specialty chemicals, and interior architectural products. In 1993, Ray Anderson, the Company's Chairman of the Board, President, and CEO, decided that the environmental performance of his industry and his company was no longer acceptable. Every year Americans discard 920 million square yards of carpet, sending 3.5 billion pounds of material to landfills.

Ray Anderson and his team have embraced industrial ecology as the framework through which Interface will reverse this pattern of waste and attain its goal of becoming a sustainable corporation by the year 2000. Interface has redesigned its business from the ground up, shifting its focus from selling carpet as a product to selling the services provided by carpets and other furnishing. At the forefront of this thinking is the "Evergreen Lease," ion which the customer purchases continually renewed furnishings, with worn modules being recycled, ultimately into new carpet.

Interface's strategy to achieve sustainability involves redesigning every aspect of the business, starting with the design of the products the company manufactures, including production processes, supplier relationships, construction, and transportation.

cutting wastes and reusing resources to create a context for choosing among technologies. An independent technology evaluation and strategic planning firm using IE methods could fill a valuable role in the waste market. Parallel opportunities could be developed in supporting industrial plant managers and R&D teams.

Industrial ecologists could be of particular use in assessing biotechnology innovations, particularly genetically engineered products. For instance, biotech firms have developed new tree "designs" for the pulp and paper industry. The promised benefits are increased resistance to pests and fiber structure requiring less energy to process. Adoption of these new breeds should be evaluated in a whole systems context, considering alternative feedstocks and the whole production cycle.

Similarly, manufacturers of agricultural supplies (such as Monsanto) are developing new plant breeds through genetic engineering. This R&D seeks higher productivity, longer shelf-life, and resistance to plant pests through built-in pesticide-like or repellent characteristics. As with forestry, such innovations in agriculture need to be evaluated in terms of the broader system of sustainable agriculture, not as stand-alone technological "miracles." This creates a significant opportunity for technology evaluation firms able to function from an industrial ecology foundation.

New Materials

New or established materials providers will discover new markets for feedstocks meeting several basic environmental requirements. These materials should be an optimal combination of the following characteristics:

- Composed from renewable and/or recycled resources
- More durable
- Recyclable
- Nonpolluting
- Biodegradable
- Require lower energy in production and use
- Enable greater strength and lower weight and volume in the final product.

Industrial metabolism, design for environment, life-cycle assessment, and input-output analysis could support achieving an optimal mix of these characteristics and integration of them with other technical requirements.

Examples of some newer materials that meet some of the technical requirements include industrial enzymes *(see sidebar on Novo Nordisk)*, biopolymers, fiber-reinforced composites, high performance ceramics, and extra-strength concrete.

Biopolymers have strong potential for replacing many synthetic plastics derived from nonrenewable, mostly petroleum sources. They may be produced directly through biological processes from microorganisms, plants and animals; or they may be synthesized chemically from such biological starting materials. They can be used in a wide variety of applications, including adhesives, absorbents, lubricants, soil conditioners, cosmetics, drug delivery vehicles, textiles, and high-strength structural materials (U.S. Congress, Office of Technology Assessment 1993). Companies embracing IE strategies will be good markets for such new materials.

New Manufacturing Processes and Equipment

Japanese researchers in the Agency of Industrial Science and Technology have offered a logical analysis of the technologies needed to achieve the closed-loop system ideal. Their report on the Ecofactory appears to be an independent development of an industrial ecology model. The model integrates design of production systems technology—including designs for the environment at product and process levels—with disassembling, reuse, and materials recycling technologies. These two large components are then linked to control and assessment technology.

The Ecofactory model describes a total system design for closed-loop manufacturing, offering the most detailed technical R&D agenda for industrial ecology to appear. The wish list includes items in energy, design, production, robotics, materials, systems, and information technologies.

Specifics such as robotics for disassembly and sorting, materials recognition systems, information systems for concurrent engineering, and many others clearly add up to major business opportunities for those who create and apply them. Companies that pioneer in this field will

Novo Nordisk has captured over 50 percent of the billion dollar world market for industrial enzymes. This competitive advantage came from early recognition that biological substitutes for synthetic chemicals would be in high demand. "We're trying to find natural solutions to industrial problems," explains Lisbeth Anker, who heads Novo's worldwide microbe search. "In the soil of an Indonesian Monkey Temple, the father of a Novo scientist unearthed an enzyme that is now widely used by soft-drink suppliers to change starch into sugar. In a pile of leaves at a Copenhagen cemetery, a researcher picked up a bug that produces an enzyme that can be used in detergents to help remove protein stains."

Enzymes are natural catalysts that speed up chemical reactions without being consumed in the process. They are biodegradable. They function best in mild conditions so their use requires up to a third less energy than many synthetic chemicals need. They are used in detergent, fabric, food processing, pulp and paper, leather, industrial cleaning, and agricultural applications.

(*Business Week* 1994)

realize profound cost savings in their own production and can open global markets for these new technologies (JETRO 1992).

SUPPORTING THE ORGANIZATIONAL CHANGE, TECHNICAL, AND INFORMATION NEEDS OF AN IE-BASED ECONOMY

IE will create new demands for the services of management and environmental consulting companies, training firms, print and electronic publishers, information system providers, and educational institutions.

Organization design: Private and public sector organizations may need to redesign their organizational structures in order to make the transition. In meeting this need, IE would be complementary to the emerging efforts in strategic environmental management. This opportunity includes consulting support for building a corporate mission, vision and strategy based in awareness of opportunities disclosed by the IE perspective. It could also require more specific changes: financial and management accounting and other systems as well as employee training and education. *See the discussion on Organizational Design in Chapter 3.*

Consulting firms able to link this level of work with capabilities in basic IE methods such as industrial metabolism, input-ouput modeling, and design for environment are likely to gain competitive advantage over those with a narrower focus. Their teams could also be strengthened by including systems ecologists.

Public policy and facilities: Governmental agencies will require support in policy and program development and in the operation of public facilities. Local or state economic development and public works agencies can benefit from IE in creating industrial development strategies and in designing more resource efficient infrastructure projects. Strategic alliances between economic development and engineering firms could meet these needs.

Another area of consulting would be creative conflict resolution on the common ground created by IE's principles, strategies, and objectives.

Product design services: Corporations are outsourcing many aspects of product design to independent design consultancies. Firms with expertise in life-cycle assessment and design for environment could use this trend to market their services. Integration of these tools with computer assisted design and rapid prototyping capabilities would be essential.

Information: Industrial ecology demands abundant flows of data as well as elegant means of turning data into information useful to a variety of stakeholders. *See the organizational design section of Chapter 3.* These needs will create new markets for a wide variety of information providers, hardware and software companies, and systems integrators.

Particularly important will be sophisticated graphic displays and utilization of innovations like neural net programming, fuzzy logic, and artificial life. The latter tools have demonstrated success in supporting brokers in perceiving and responding to the complexities of financial markets. They could do the same for industrial ecologists.

Finally, companies, communities, and industrial parks will need consulting, software, and information tools to support the design and operation of closed loop systems among companies (industrial ecosystems). This could include environmental monitoring and information systems and independent auditing of environmental performance, providing feedback to individual companies, the community, and agencies. This service could also provide third party reporting that is required in many voluntary programs.

INTEGRATING TECHNOLOGIES AND METHODS INTO INNOVATIVE NEW SYSTEMS

A foundation in industrial ecology could enable established larger corporations, entrepreneurs, or joint ventures of smaller companies to seize competitive advantage through systemic design. The key would be assembling diverse elements into innovative new systems that meet customer needs while dramatically reducing environmental impacts.

For example, a major construction and engineering company (or a consortium of smaller firms) could design and build a decentralized energy infrastructure in a country in the process of industrializing. The system might integrate modular design, a mix of smaller fossil fuel plants and renewable sources, and a strategy of highly efficient use and reuse of energy (especially waste heat).

The strength of the strategy would be its integration of existing and new elements into a fundamentally new system. The country could meet its energy needs in a phased development with much lower environmental impact than that of a traditional large-scale solution. The company could effectively compete against much larger and less agile corporations. *See the scenario in Chapter 9.*

A related business opportunity would offer support to the integration of industrial facility design. This might be a service of an architecture and engineering firm, a construction firm, or a smaller consulting company. The objective would be a plant in which all aspects of design were interrelated in a whole system.

Integration would include the environmental, financial, and technical aspects of the design of the building, the industrial processes taking place in it, its landscaping, and its relationships to local and regional ecosystems and to its neighbors. The design would also seek optimal solutions across the whole life cycle of the building, from planning through construction and commissioning, to operation, and final decommissioning.

At a more ambitious level, IE suggests basing a company's mission on delivery of high quality service rather than material products. In this vision, the company would own the products that deliver the service and optimize their design for a long life. The equipment would be durable, modular, easily repaired or rebuilt, multifunctional, and often integrated into a whole system for service delivery.

Some copier manufacturers have already pioneered this approach. A hypothetical example would be for an appliance manufacturer to offer integrated household cooling, heating, lighting, and cooking systems on a lease or fee basis. *See Chapter 4 on Product-Life Extension and The Service Economy for a discussion of this concept.*

An example of an IE project for the public sector would be planning infrastructure projects with a strong emphasis on programs and services that enable more efficient use of a landfill or water system. The higher efficiency could reduce the size of the construction investment.

RESOURCES ON BUSINESS OPPORTUNITIES

CADDET, http://www.ornl.gov/CADDET. A database of demonstration projects on energy efficient and renewable energy technologies.

COMPENDEX, telnet://192.204.252.2. Contains abstracts and indexing to some 425 international journals and key conference proceedings from 1987 to the present. Its broad subject coverage includes chemical engineering, civil engineering, metals and mining, manufacturing engineering, and more.

DOE Efficiency and Renewable Energy Network (EREN), http://www.eren.doe.gov. Contains information on a large variety of energy-related recycling and conservation technologies. Also includes published searches on the COMPENDEX and patent database which can be purchased from NTIS.

DOE Environmental Inventions and Innovations, http://www.nttc.edu. Lists industrial and environmental technologies sponsored by the Departments of Energy and Commerce. The citations include a description of the technology, a contact person, and the technology's stage of development.

The EnviroLink Network, Suite 236, Hamburg Hall, 5000 Forbes Ave., Pittsburgh, PA 15213. Telephone: 412-268-7187, Fax: (412) 268-7036, e-mail: admin@envirolink.org. The Environmental World Wide Web Listing at EnviroLink lists most environmental World Wide Web services. A full listing of all environmental computer networks, including gopher, ftp, etc., is available via the EnviroWeb, a project of the EnviroLink Network.

Federal Laboratory Consortium for Technology Transfer, 317 Madison Ave., Suite 921, New York, NY 10017-5391. Telephone: 212-490-3999, Fax: 986-7864. Promotes the transfer of federal R&D capabilities into the private sector and state and local governments. Newsletter is NewsLink.

Investment Recovery Association, 5818 Reeds Rd., Mission, KS 66202-2740. Telephone: 913-262-4597, Fax: 913-262-0174.

The National Environmental Technology Applications Corp. (NETAC) at the University of Pittsburgh provides a toll-free information hotline service to assist environmental technology developers in their commercialization efforts. The hotline (1-800-48NETAC) offers information about the commercialization process, public and private financing sources for technology development, commercialization, and related issues. For more information: Celeste E. Steffen. NETAC, University of Pittsburgh Applied Research Center, 615 William Pitt Way, Pittsburgh, PA 15238. Telephone: 412-826-5511.

National Technical Information Service, 5285 Port Royal Rd., Springfield, VA 22161. Telephone: 703-487-4650.

Office of Scientific and Technical Information, U.S. Department of Energy. 1994. *The Office of Technology Development Technical Reports. A Bibliography.* September 1994. This bibliography contains information on scientific and technical reports sponsored

Phoenix Designs is a Herman Miller subsidiary manufacturing office systems and furniture, largely from recycled materials. Designers have applied the company's environmental values to planning a 590,000 square foot manufacturing and office building.

The Phoenix Designs building will be closely integrated into its ecosystem, which is a transition between oak/hickory and beech/maple habitats. Landscaping will feature native trees, shrubs, grasses, and wildflowers. The pattern of reforestation will shelter the new building from winter winds and snow drift. Existing and new wetlands will provide wildlife habitats and cleansing of stormwater runoff before it goes into the adjacent Black River. (Runoff from paved areas will first go into settlement basins to remove petroleum or metals.) Exercise and nature paths for employees will wind through the layered site, prepared with minimal impact on natural forms. The landscape will "enter" the building itself through plantings in courtyards and along an internal skylit street.

(William McDonough + Partners and Herman Miller Inc. 1994)

by the Office of Environmental Management from its inception in 1989 through June 1994. Future issues contain reports from technology development activities and will be published biannually. This bibliography can be accessed through Dialog and ITIS.

Other World Wide Web Sites of Interest

Center for Clean Technology (CCT) at UCLA: http://cct.seas.ucla.edu/cct.pp.html.

Center for Green Design and Manufacturing: http://euler.berkeley.edu/green/cgdm.html.

Enviroene: http://wastenot.inel.gov:80/envirosense/.

Recyler's World: http://granite.sentex.net:80/rec.

U.N. Environment Program—Sustainable Product Development: http://unep.frw.uva.nl/.

Renewable Energy Organizations

Energy Research Institute, 6850 Rattlesnake Hammock Rd., Hwy 951, Naples, FL 33962. Telephone: 813-793-1922, Fax: 813-793-1260. Individuals and companies interested in alternative energy sources. The goal is to achieve energy independence through development of wind generators, methane digesters, solar cells and collectors, steam generators, and hydrogen production with special focus on alcohol production for fuel purposes.

Fuel Cell Association, P.O. Box 66392, Washington, DC 20035-6392. Telephone: 301-681-3531, Fax: 301-681-4896. Fuel cell researchers, developers, designers, manufacturers, installers, and system users. For the promotion, development, and use of fuel cells, which are electrochemical devices and produce electricity directly from hydrogen or hydrocarbon fuel and an oxidant. They are presently considered to be the source with the highest energy conversion rate for commercial, industrial and residential uses.

Passive Solar Industry Council, 1511 K Street NW, Ste. 600, Washington, DC 20005. Telephone: 202-628-7400.

Renewable Fuels Association, 1 Massachusetts Ave. NW, Ste. 820, Washington, DC 20001. Engineering and financial firms, marketers, producers, and state governments promoting fuel technologies, mainly alcohol fuels. Represents renewable fuels industry to the government and monitors every phase of biomass fuel production.

Solar Energy Industries Association, 777 North Capitol St. NE, Ste. 805, Washington, DC 20002. Telephone: 202-408-0664, Fax: 202-508-8536. Companies, universities, and utilities. Involved in research and education.

Utility Photovoltaic Group, 1101 Connecticut Ave NW, Ste. 901, Washington, DC 20036. Telephone: 202-857-0898, Fax: 202-223-5537.

Recycling Associations

Center for Waste Reduction Technologies, 345 E 47th St., New York, NY 10017. Telephone: 212-705-7407, Fax: 212-705-3297. Businesses, corporations, academicians, etc. Research; educational programs.

Global Recycling Network (GRN), 2715A Montauk Highway, Brookhaven, NY 11719, USA. Telephone: 516-286-5580, Fax: 516-286-5551. An information service set up on the Internet to help businesses worldwide exchange leads to trade recycled resources, surplus goods, or used machinery. To receive information automatically, E-mail a message to infor@grn.com; for fax-back information, dial 1-516-286-3686 on your fax and press "Start" after you hear the fax tone.

Industry Council for Electronic Equipment Recycling, 6 Bath Place, Rivington St., London EC2 4A, UK. Contact Graham Marson, Boots (electronics mfr.), Telephone: 44-602-866-671, Fax: 44-602-860-695. Boots, British Telecom, and computer manufacturers, including manufacturers of raw materials and equipment, retailers, and distributors. Recycling information; encouragement of market development for recycled goods; development of infrastructure for genuine recycling.

Institute of Scrap Recycling Industries, 1325 G St. NW, Ste. 1000, Washington, DC 20005. Telephone: 202-466-4050, Fax: 202-775-9109. Processors, brokers, and consumers engaged in recycling ferrous, nonferrous, and nonmetallic scrap; related industry organizations. Specialized education and research programs.

Investment Recovery Association, 5818 Reeds Road, Mission, KS 66202-2740. Telephone: 913-262-4597, Fax: 913-262-0174. Investment Recovery is the process of recycling, reusing, and selling the surplus materials, equipment, by-products, and wastes acquired or generated by a company in conducting business. The Association includes representatives of over 100 corporations seeking the highest value reuse of their surplus materials and machinery.

National Materials Exchange Network Web home page: http://www.earthcycle.com/nmen/. The National Materials Exchange (NME) Network makes it possible to search for and identify materials and goods that are available, as well as environmentally focused services. The Network maintains two databases, the NME Materials and Goods Database and the NME Services Database. Storage, disposal, transportation, and consulting are examples of the types of services listed. An account is required to search the databases at NME, but access is free and the account can be immediately requested and activated online.

National Recycling Coalition, 1101 30th Street, NW, Ste. 305, Washington, DC 20007-3708. Telephone: 202-625-6406, Fax: 202-625-6409. The Recycling Technology Assistance Partnership (ReTAP) is a nationwide technology extension program that is working to advance industry's use of recovered and recycled materials. Among the products and services ReTAP offers is the Technology Service Tool Kit, which provides information on how to design and implement recycling technology programs that can effectively meet the needs of local manufacturers. The Tool Kit will consist of a series of specific procedures, methodologies, and related planning and implementation suggestions. The ReTAP National Network is also offering a series of one-day regional workshops on planning and implementing technology services that will be tailored to meet the needs of the participants, using local examples.

Bibliographic Resources

Business and the Environment. Cutter Information Corporation, 37 Broadway, Suite 1, Arlington, MA 02174-5552. Telephone: 617-641-5125, Fax: 617-648-1950, E-mail: Editor—bate@igc.apc.org. Client Services: dcrowley@cutter.com.

Business Week. 1994. "Novo Nordisk's Mean Green Machine—It's at the Top of a Growing Market: Finding Natural Substances to Replace Chemicals." November 14.

FitzGerald, Chris. 1995. *Environmental Management Information Systems*. FitzGerald shows how to develop effective EMIS at the plant, division, and corporate levels. The book provides guidelines for selecting EMIS, covering project planning and justification, requirements, specifications, evaluation criteria, commercial software products, and project life cycles and implementation. McGraw-Hill, Inc. Blacklick, OH.

JETRO (Japan External Trade Organization). 1992. "Eco-factory—Concept and R&D Themes." Special issue, *New Technology*. Tokyo. Report based on work of the Ecofactory Research Group of the Mechanical Engineering Laboratory, Agency of Industrial Science and Technology.

Martin, Sheila, et al. 1995. "Technologies Supporting Eco-Industrial Parks." In *Developing an Eco-Industrial Park: Supporting Research*, Volume 1, Final Report, Research Triangle Institute Project Number 6050, Research Triangle Park, NC.

McDonough, William + Partners and Herman Miller Inc. 1994. *The Phoenix Designs Project*. Charlottesville VA.

Phillips Petroleum. No date. *Cleaner Earth, Investment Recovery and Recycling*. Company brochure. Tulsa, OK.

Romm, Joseph J. 1995. *Lean and Clean Management: How to Boost Profits and Productivity by Reducing Pollution*. Kodansha International, NY. Case studies and guidelines for systemic application of pollution prevention and energy efficiency.

U.S. Congress Office of Technology Assessment. 1993. *Biopolymers: Making Materials Nature's Way*. September. OTA-BP-E-102, Washington, DC. Comprehensive background paper on the potential for substituting biologically based materials for many polluting chemically based polymers.

Watanabe, Chihiro. 1993. "Energy and Environmental Technologies in Sustainable Development: A View from Japan," *The Bridge*. National Academy of Engineering, Washington, DC, Summer. Sketches Japanese work in industrial ecology beginning in 1972.

A FUTURE SCENARIO: An Insurance Company Assesses its Investment Portfolio

In the last half of the 1990s, several prominent insurance companies instituted a major examination of their investment strategies, using insights and tools from industrial ecology. Catastrophic claims from hurricanes, flooding, and other weather events were causing very high payouts from the industry ($39.5 billion for U.S. companies alone between 1989 and 1992). Insurance executives began to explore the possibility that these major losses could be connected to the impact of global warming on climate patterns. They were also concerned about liabilities they were covering in toxic spills and polluted industrial sites.

An industrial ecology consulting group proposed analyzing insurance investment portfolios, reviewing their total environmental impact, not just the issues of climate change. Gulf Reinvestment (GR), a Florida based company, contracted for a pilot project. Work began in mid-1999 with an executive briefing and a more intensive workshop for staff investment and risk analysis researchers. Initially, the idea of an insurance company, its industrial clients, and its investment portfolio being a living system seemed quite alien. Discussion of IE's analytic tools demonstrated that the idea can benefit even a financial institution. An industrial ecology review of one of GR's major insurees with heavy Superfund exposure helped bring the concept home.

With a common understanding of IE's potential, the consultants and corporate researchers worked together to analyze major inputs and outputs of the company's 15 most energy and materials intensive clients and investments (using standard industry and regulatory data). The pilot study of the portfolio's industrial metabolism traced the following:

- Drawn upon nonrenewable resources
- Greenhouse gas emissions
- Other pollution to air, water, and land
- Generation of solid wastes
- Patterns of energy use.

The team went on to review possible impacts of these findings on the insurance company's risk exposure. This quick and dirty pilot study indicated some critical points where GR supported major contributors to greenhouse gases and other global environmental effects. Environmentalists had been demanding a strategy of disinvestment in fossil fuel companies. Gulf executives insisted that this would be a quick-fix with fiduciary as well as environmental hazards. They called for a truly systemic approach. They also saw that they would have to build strategies for risk reduction through collaboration with other major players in the insurance industry, as well as with their major insurees and with the companies in which Gulf invested.

In early 1998, GR decided to issue its IE team's initial findings as the Gulf Reinvestment Environmental Report Card. This report was used to stir dialogue within the company, with its insurees, and with the firms in GR's portfolio. The communications package included a print report, a video tape, an overview of industrial ecology, and a computer disk with a survey for each company in its portfolio. The latter raised questions about environmental performance objectives, strategies for reducing greenhouse gas and acid rain emissions, and the status of programs for implementing these strategies.

This process continued through 1998, leading the majority of Gulf's clients and investments to set voluntary goals for reductions in emissions greater than those the U.S. EPA was suggesting.

A new cluster of hurricanes and floods in the summer and fall prompted Gulf Reinsurance and other major reinsurance companies to organize a series of informal CEO level retreats with leaders in the energy industry. Gulf bluntly asked how could the insurance industry support the energy industry's achievement of much higher energy efficiency, diversification into renewable sources, and the phasing down of use of fossil fuels.

In early 1999, GR started analyzing the economic and environmental impacts of a strategy of technological and business diversification in the energy industry. Its research team drew upon another industrial ecology tool—dynamic input/output (I/O) modeling *(see Chapter 3)*. I/O models enabled them to compare the impacts of a business as usual scenario and an energy diversification scenario. They also factored in growth in the environmental and renewable energy venture funds Gulf had set up to finance commercialization of demonstrated new technologies. The resulting what-if tool continued to play a major role in the company's portfolio analysis and its ongoing dialogue with major companies in which it invested or whom it insured.

Gulf offered notable leadership to its industry and to major firms in the industry's portfolio. Without ever threatening disinvestment, the company supported a more rapid reduction in greenhouse gases and other emissions than any corporations thought possible at the beginning of the process. An insurance company, acting in terms of its economic self-interest, helped define the common self-interest in sustainability.

Afterword: The Present Factual Background for This Future Scenario

Beginning in the early 90s, scientists of the environmental action organization, Greenpeace, made a case for the role of global climate change in triggering more frequent hurricanes and floods and record claims on insurance companies. (Leggett, Jeremy. 1993. Climate Change and the Insurance Industry. Greenpeace, London.)

In 1995 and 1996, insurance executives attended the Berlin Climate conference with great interest. Frank Nutter, President of the Reinsurance Association of America, said, "It is the threat to people from natural events that drives much of the demand for insurance; yet it is the same threat of future natural catastrophes that could jeopardize the industry's financial viability." He pointed out that in the U.S. 21 out of the 25 largest catastrophes (in terms of insurance claims paid) had occurred between 1985 and 1995. Sixteen of these involved wind and water, mainly in the form of hurricanes. (*Christian Science Monitor.* March 29, 1995. "Insurance Firms Ask If Global Warming Swells Disaster Rate." Boston, MA.)

Richard Keeling, a former Lloyds of London executive at the Berlin meeting, said, "Every major economy in the world where we have significant exposure has had a loss . . . we started getting concerned about this problem, and we got our experts to look at the reasons for these losses. They turned around and said 'Well, we can't prove that we have a definite global warming problem, but by the time we can, you chaps are in real trouble.' " ("Insurers Launch Joint Effort to Tackle Environmental Risks." April 7, 1995. *Environment Watch: Western Europe*. Vol. 4, No. 7. Cutter Information Corp., Arlington, MA.)

Five major European insurance companies have begun an effort to take their customers' environmental performance into account when setting premiums and to improve their own environmental record. Senior executives of UNI Storebrand (Norway), Swiss Re, Gerling Global Konzern (Germany), General Accident (UK), and National Provident Institution (UK) pledged to integrate environmental considerations and a "precautionary principle" into their business goals.

Åge Korsvold, President of UNI Storebrand said, "Our objectives as an insurance company and the society's objectives for a sustainable development are closely linked. From an economic viewpoint, it is more profitable to employ resources in prevention than to pay insurance claims. Environmental risks are also financial risks."

Munich Reinsurance Company, the world's largest reinsurer, had announced prior to the summit that it was raising its estimates of costs caused by natural disasters because of their increasing frequency and said that it believed this increase and the risk in global mean temperatures was no coincidence ("Insurance Companies Launch Environmental Initiative," *Business and the Environment,* May 1995. Cutter Information Corp., Arlington, MA, p. 9).

6 Industrial Ecology Opens New Opportunities for Government

INTRODUCTION

A core question for government, in an era of shrinking budgets, is how to make the transition to a sustainable industrial economy. Such an economy would meet the needs of still growing human populations without continuing to deplete non-renewable resources and degrade local and global environments. The broad mission we face has three primary components:

- Modifying human activities to match global, regional, and local ecological constraints
- Preserving, restoring, and renewing natural systems
- Maintaining viable economies and industries to meet growing human needs during the transition.

Government can play a role in supporting these activities by, for example, defining policy and regulations, organizing agencies to work more effectively in partnership with the private sector, guiding and financing research and development, and facilitating sustainable economic development.

Several measures of sustainability developed in an IE context are useful guides to government action across all of these areas.

Measures of sustainability suggested by industrial metabolism (see Chapter 3).

1. *Ratio of virgin to recycled materials*
2. *Ratio of actual/ potential recycled materials:* ratio between volume of materials that could be recycled to the fraction actually recycled
3. *Ratio of renewable/fossil fuel sources*
4. *Energy productivity*
5. *Materials productivity: The economic output per unit of material input*
6. *Resource input per unit of end-user service:* This ratio assesses resource use against the useful function gained and maintained for the end-user.

Two of the criteria for sustainability suggested by the Natural Step add important elements beyond resource productivity (see Chapter 2 and Holmberg 1995).

- *Does the action reduce the use of long-lived synthetic products or molecules?*
- *Does it preserve or increase natural diversity and the capacity of ecological cycles?*

The Fieldbook for Development of Eco-Industrial Parks *(Lowe et al.1995) outlines an environmental performance assessment framework based upon IE (see Chapter 2, Guidance From Nature in Setting Objectives).*

In addition, governments are directly responsible for implementing sustainable approaches to the design and operation of public facilities and infrastructure.

Industrial ecology can support the work of public policy makers, researchers, and regulators who are seeking more effective, systemic approaches to environmental protection and economic development because

- Its global view offers an appropriate context for prioritizing risks and identifying points of high leverage for change.
- IE methods, such as industrial metabolism and dynamic input-output modeling, provide means for assessing alternative policy options.
- Organizational design could benefit from IE's strategy of learning from the dynamics and principles of ecosystems, particularly their processes of regulation and self-regulation.
- IE can assist policy making and research funding in energy, transportation, agriculture, and service industries as well as in manufacturing.
- Public sector facilities and systems managers can benefit by applying industrial ecology principles in their operations. These range from military bases to local water, sewage, and solid waste systems.
- IE can help economic development policy makers and managers form more effective, sustainable strategies.
- Implementation of IE in government facilities opens many business opportunities through demand for new technologies and processes and new applications of old ones.

There are specific opportunities for implementing IE in four areas of government activity: policy and regulation, research and development, procurement, and local economic development.

POLICY AND REGULATION

Industrial ecology is often congruent with emerging changes in policy and regulation. Many creative initiatives are under way, ranging from Technologies for a Sustainable Future (U.S. Office of Science and Technology Policy) to the Common Sense Initiative (US-EPA) to Industries of the Future (Department of Energy).

IE's emphasis on systemic approaches supports the move upstream from end-of-pipe controls and the trend toward multi-media permitting and enforcement. Viewing industry as a proactive participant in the improvement of environmental performance fits well with the development of voluntary programs, economic incentives, and cooperative setting of objectives. IE's methods of analysis provide means for prioritizing risks, weighing trade-offs, and setting objectives. IE's global view is a valuable context for integrating these and other currents in regulatory reform.

Is There a Conflict Between Environmental Regulation and Economic Performance?

There is growing evidence that environmental regulation can actually benefit competitiveness. Research by Michael Porter and Claas van der Linde[2] links the economic value of regulation to resource productivity, a basic industrial ecology measure of sustainability. They argue that a static view of the role of regulations is the basis of the perception of conflict between ecological and economic needs. This old view insists that there is an inevitable trade-off between the social benefits of mandated environmental standards and industry's private costs of control, prevention, and cleanup. This static model has largely been shared by industry and government and is closely linked to the traditionally adversarial nature of the relationship between the two (Porter and van der Linde 1995A).

The authors argue that this simple analysis overlooks the dynamic character of industrial innovation in response to external pressures. They offer case studies and statistical evidence indicating that companies in the U.S. and Europe are gaining competitive advantage through the higher

"Properly designed environmental standards can trigger innovations that lower the total cost of a product or improve its value. Such innovations allow companies to use a range of inputs more productively—from raw materials to energy to labor—thus offsetting the costs of improving environmental impact and ending the stalemate. Ultimately this enhanced resource productivity[1] makes companies more competitive, not less."

(Porter and van der Linde 1995A, p. 120)

[1]The concept of resource productivity is the basis of several measures of sustainability suggested in the industrial metabolism work of Robert Ayres. *See Chapter 3.*

[2]Porter is at Harvard Business School and van der Linde at the International Management Research institute of St. Gallen University in Switzerland.

resource productivity created by their responses to regulatory pressures.

Their dynamic model for assessing the impacts of regulations includes not just the costs of compliance but also the opportunity costs of pollution/inefficiency ("wasted resources, wasted effort, and diminished product value to the customer"). When the companies act upon the opportunities, they save significantly from their investments in technical changes that improve environmental and economic performance simultaneously.

By eliminating inefficiencies in the use of resources all along a product's life cycle, managers cut costs and create new values. These inefficiencies include incomplete utilization of material and energy resources, poor process controls, product defects, storage of wastes, discarded packaging, costs of products to customers of pollution or low energy efficiency, and the ultimate loss of resources through disposal and dissipative use. Poor resource productivity also triggers the costs of waste disposal and regulatory penalties.

Porter and van der Linde discovered a wide range of innovations to improve resource productivity:

- Simplified product design and packaging
- Recycling scrap
- Improved secondary treatment
- More efficient utilization of inputs
- Substitution of nonpolluting input materials that often cost less
- Elimination of regulated input materials through process changes
- Improved process consistency
- Processing toxic wastes, emissions, and non-hazardous wastes into usable forms
- Recovery of valuable materials from product recycling and takeback.

Benefits include lower product costs, reduced downtime, better product yields, and improved product performance. For example, a California Dow Chemical plant spent $250,000 to redesign one production process to eliminate the use of a regulated material, caustic soda. Total annual savings from the changes are $2.4 million.

This study focused on one important objective, improving the productivity of resource use. The driver was regulation of releases, not resource use, but the companies adjusted their production systems to gain both the mandated performance and the economic benefits. From an IE perspective, the plants gave priority to whole system solutions, not end-of-pipe controls. This systems approach is preferable to simply searching for customers to buy wastes as by-products. The exchange of material or energy by-products is an important option for companies, but it should be chosen only after a systems analysis shows it to be the preferred solution.

Reform of Regulations and IE

Present regulatory barriers to IE

The existing laws have some important barriers to implementing basic IE strategies:

- The process of developing and implementing environmental law is fragmented across Congressional committees, agencies, media of pollution, and stages in a product's life cycle. This fragmentation means there is no coherent process for setting priorities in a whole systems context.
- Industrial ecologists encourage adoption of the principle that all "waste" energy and materials be seen as potential products. Regulations, on the other hand, rigorously define wastes and their sources in ways that at times severely limit their reuse or recycling.
- Consumer protection laws require that products containing any reclaimed or remanufactured parts be labeled "used." This discourages recycling of components or subassemblies.
- Federal Superfund statutes and related state law make the clean-up and reuse of contaminated industrial land difficult and costly.
- Government procurement specifications or standards often favor use of virgin materials, making it difficult to use recycled ones.

- Antitrust laws may discourage the vertical integration that may be needed for manufacturer take-back of products, management of life-cycle issues, and flows of by-products.
- Laws directly relating to biodiversity and the health of ecosystems are fragmented and not well connected to either systems of production or consumption.

IE tools of systems analysis and design—such as industrial metabolism studies, life-cycle assessment, and design for environment—help overcome the fragmentation by medium of pollution or stage of a product's life cycle. Getting past some of these barriers simply requires empirical evidence that innovations such as industrial by-product exchange are environmentally safe and economically feasible. In general, industrial ecology will support the creation of policies that are open to adaptation as new knowledge of both technology and the environment evolves. (See Powers and Chertow 1997 and Martin 1996 for in-depth analysis of regulatory barriers to IE and insights into the changes required.)

Reinventing regulations

Regulations designed in terms of the dynamic, systems view Porter and van der Linde propose would enable industry to innovate, use resources more productively, and enhance competitiveness. The principles they propose parallel central themes developing in environmental agencies and being expressed in new initiatives. All would support application of industrial ecology approaches. In turn, industrial ecology would assist the process of regulatory design. Some of these themes include:

Focus on outcomes, not technologies: Prescription of technical solutions has been one of the most limiting aspects of environmental regulation. If industry is not locked into a given answer it has the room to innovate, and to achieve or exceed performance targets set through a collaborative process. With this freedom, firms are likely to save money while benefiting the environment, as Porter and van der Linde's research establishes.

Make the regulatory process more stable and predictable: Industry needs to be able to plan investments in product development, technology, marketing, and organizational support. Regulatory surprises, even including swings into deregulation, are often quite damaging to competitiveness.

Require industry participation in setting standards from the beginning: This principle is rapidly becoming the common wisdom of regulators. Within a clear public process for setting performance objectives, business managers and technical staff should be able to apply their knowledge and skills to create the means of achieving those objectives.

Current programs like the Common Sense Initiative (CSI) reflect many of the above principles as well as others important to the implementation of IE. The CSI is working to integrate across media of pollution (air, water, and solid waste) with six industries (the iron and steel industry, the electronics and computer industry, the metal plating and finishing industry, the auto industry, the printing industry, and the oil refining industry[3]).

In each industry EPA is collaborating with state and local regulators and citizen groups to set performance goals within which the companies have flexibility for designing the means of achieving the goals. The CSI also will seek streamlining of environmental permitting and simpler reporting. It moves regulation to a more systemic style by focusing on overall environmental performance by industry, by integrating concerns across media and levels of government, and by recognizing that industry has the knowledge needed to define the most effective means of achieving publicly defined performance goals.

Other reform initiatives are working on an issue important to industrial ecologists: making it easier for plants to utilize and trade wastes as by-products. In IE "There are no wastes, only residues that should be designed so that an economic use can be found for them" (Graedel and Allenby 1995, p. 83). Without undercutting present safeguards on handling by-products, regulations such as RCRA need to be rewritten to enable storing, selling, transporting, and brokering them. US-EPA has created an Office of Regulatory Reinvention to administer CSI, Project XL, and other voluntary programs.

The science of regulation: Industrial ecology is a specific application of systems science, one particularly relevant to

"I draw three simple but important lessons from the past 25 years. One, we must recognize that nature is a system. We must recognize the integration of our air, our water, and our land. Merely regulating on a pollutant-by-pollutant basis is not enough. Two, we must change the process. We must move beyond an adversarial process. We must inform and involve those who must live with the decisions we make—the communities, the industries, the people of this country. And three, we have not finished the job. We must preserve and strengthen the principles of environmental protection, while changing the means by which we achieve these protections.

(Browner 1995)

[3]These industries make up nearly 15 percent of U.S. gross domestic product. They account for 345 million pounds of toxic releases—1/8 of all toxic emissions.

the content involved in reinventing regulations and policy. Cybernetics is another branch of systems study fundamental to the processes involved in regulatory redesign. One of its founders, Norbert Weiner, defined it as "... the science of communication and control in animals and machines." Design of regulations can benefit from this science of regulation. Cybernetics offers insights into design of organizational and inter-organizational structures and channels for communication and feedback. Its methods for enhancing self-regulatory behavior are particularly important, given the present trend toward partnership between regulators and the industries they seek to regulate. The resources at the end of this chapter include some basic works on cybernetics.

In the last decade, the study of complex adaptive systems has added a new level of systems science that is also very applicable to development of environmental policy and regulations. Researchers in this field are developing concepts and methods for managing systems that are non-linear, that have high variety, and whose future states cannot be readily projected through analysis of present trends. Global climate change, ecosystem dynamics, market economics, and synergistic interactions among chemicals are examples of such complex adaptive systems (Holland 1996).

Beyond Risk Assessment

Prioritization of risks is a fundamental requirement for development of environmental policy, one that affects application of IE tools (like design for environment), design of regulations, scientific research agendas, and private or public investment in technology R&D and implementation. At the same time, IE methods can support the process of prioritization. Some elements of a system for prioritization of human and ecological risks include:

- Spatial scale of impact
- Importance of the ecosystem that is exposed
- Severity of the hazard
- Intensity of exposure (in terms of population involved and degree of exposure)
- Economic impacts
- Time scale of effect and ecological recovery.

From an industrial ecology perspective it is vitally important to remember that prioritization of risks is not hard science, but a dialogue between scientific research and public values. Science can tell us only the present state of knowledge about the relative health and ecological risks of human activities, principally the use of chemicals, radioactive materials, and other substances. (This knowledge itself is very incomplete and often is soft because it deals with complex, non-linear systems and lacks "statistical power."[4]) Public choices then have to be made as to which values to use in establishing priorities. On the one side, industry argues strongly for "science-based risk assessment" and cost/benefit analysis of regulations, arguing that "lower-level" risks are not worth the cost of measures to eliminate them.

On the other hand, public interest advocates and many scientists argue that we must act with much greater caution. They say our understanding of the full scale of risks we face is fragmentary, and there are clear limits to the power of science to define risk levels. Emerging issues, such as endocrine disrupters and synergistic interactions among combinations of substances, could undermine the foundations of risk assessment. If continuing research supports hypotheses suggested in these areas, regulators and industry alike may need to rethink environmental management strategies.

International pressure is already demanding a move beyond risk assessment and risk management to a more proactive, preventive, and precautionary approach. Industrial ecology offers a context for negotiating the terms of the transition.

The pressure for fundamental change in dealing with risks is represented by The International Joint Commission (IJC) the Official U.S.-Canadian body responsible for monitoring progress in achieving the objectives of the Great Lakes Water Quality Agreement (see sidebar). The IJC report says that "conventional scientific concepts of dose-response and acceptable "risk"can no longer be

". . . prioritization and reordering of environmental values, both among themselves (e.g., is Superfund, human carcinogenicity, or global climate change more important?) and in the broader context of other social values (e.g., employment, private property rights) can only be accomplished through the political process.

"While it is doubtful that an unambiguous, uncontentious prioritization of values is possible, some broader consensus is necessary to provide support for further progress: How, for example, can an engineer be expected to design a "green" product when what is environmentally preferable cannot be made clear?

"This will not be a trivial task. It will require, for example, the development of comprehensive risk assessment (CRA) methodologies, which evaluate and balance risks and possible benefits on a systems-wide basis. While such approaches have been suggested, ***no such methodologies yet exist, nor is it clear that the data or organizational structures necessary to support implementation of CRAs are currently available.****"*

(IEEE EHSC 1995)

[4]Statistical power is the probability that a given experiment or monitoring program will detect a certain size of environmental effect if it actually exists. A report to the OECD found that due to limited budgets, small samples, imprecise sampling, and a huge number of variables and range of variability in nature, statistical power is often low in environmental studies (Rolfe 1994).

The International Joint Commission (IJC) is the Official U.S.-Canadian body responsible for monitoring progress in achieving the objectives of the Great Lakes Water Quality Agreement. The IJC biennial reports have taken an aggressive stance regarding risk assessment. The IJC's 7th report (IJC 1994) calls for:

- *Phase-out ("sunsetting") of all persistent toxic substances from the Great Lakes ecosystem*
- *A ban on the manufacture and use of chlorine*
- *An end to reliance on risk assessment*
- *A ban on solid waste incineration*
- *A reversal of the policy that assumes chemicals are innocent until proven guilty*
- *Adoption of the principle of precautionary action (which says: wherever it is acknowledged that a practice could cause harm, even without conclusive scientific proof that it does cause harm, the practice should be prevented and eliminated)*
- *An end to chemical-by-chemical regulation, substituting an approach that eliminates whole classes of chemicals by strategically preventing the formation of the persistent toxic substance in the first place.*

defined as "good" scientific and management bases for defining acceptable levels of pollution. They are outmoded and inappropriate ways of thinking about persistent toxics."

While some may see this as an extreme position, the Commission bases its arguments upon the precautionary principle,[5] a values-based approach. The scientific basis for the IJC's recommendations are the early warning signs of the effects of persistent toxins that accumulate over time in ecosystems and animals, including, of course, humans. This evidence has focused upon the possible impacts of endocrine disrupters.

Endocrine Disrupters: Our Stolen Future (Colburn et al. 1996) summarized research suggesting that a variety of commonly used chemicals may be interfering with the normal functioning of human and animal endocrine systems. The hypothesis explored in this research is that these substances may be increasing the rates of selected cancers and causing reproductive system abnormalities, learning and behavioral problems, and immune system deficiencies.

With some chemicals, persistence and accumulation in the environment and in human and wildlife tissues further increase the risk factor. These substances also show, in some cases, a non-linear dose response. This means that doses of endorcrine-related substances lower than those used in toxicological studies sometimes generate higher impacts because of the unique power and dynamics of hormonal action. Current regulatory practice is to extrapolate down from the high dosage tested, assuming a linear response and sometimes inaccurately projecting a safe level of dosage (Colburn et al. 1996, pp 169-79, 205-6). Such non-linearity generates high complexity in risk assessment.

The list of suspected agents is long, including heavy metals, pesticides such as DDT and lindane (both still in use in many countries), PCBs, and many compounds used in industrial and household products or generated through waste treatment, especially incineration. The possible im-

[5]The Rio Summit Declaration defines the precautionary principle narrowly: "Where there are threats of serious or irreversible damage, lack of full scientific certainty shall not be used as a reason for postponing cost-effective measures to prevent environmental degradation." An International Conference on the North Sea defined the principle in terms of more affirmative action. The conference declaration said that potentially damaging pollution emissions should be reduced even "when there is no scientific evidence to prove a causal link between emissions and effects" (Rolfe 1994).

pacts of such substances on endocrine functions may be augmented by the action of naturally occurring plant and animal hormones.

At a policy level, Congress has started to respond to this risk with amendments contained in the Food Quality Protection Act of 1996. This legislation instructs the EPA to "determine whether certain substances may have an effect in humans that is similar to an effect produced by a naturally occurring estrogen, or such other endocrine effect as the Administrator may determine." If substances are found to have endocrine effects, the laws provide that the Administrator, "shall, as appropriate," take action under available law, "as is necessary to ensure the protection of public health."

The EPA has convened an Endocrine Disrupter Screening and Testing Advisory Committee, which has a mandate to consider both wildlife and human effects across a full range of endocrine functioning (not just estrogenic). An initial report suggests that no action is needed to protect public health or the environment but that more study is needed (Crisp 1997). On the other hand, Dr. Robert Huggett, EPA's assistant administrator for research, told the *New York Times* that the 300 studies upon which this committee's report was based "demonstrate that exposure to certain endocrine [hormone] disrupting chemicals can lead to disturbing health effects in animals, including cancer, sterility, and developmental problems" (*New York Times*, March 14, 1997). (See also Center for the Study of Environmental Endocrine Effects web site for more information.)

The Chemical Manufacturers' Association has also established a special committee to address the implications of this issue for the industry.

Synergy among toxic substances: The complexity of risk assessment becomes even higher when we consider synergistic interactions among substances. We know from medicine that even some individually healing substances can become deadly or highly damaging when prescribed in combination. In workplace safety and health management, employees are trained to understand which combinations are potentially explosive, or for which they require special protective gear. However, toxicologists normally study ecological and health risks of substances one at a time, and so far have reliable risk data on only a fraction of the synthetic chemicals in use.

In 1996, a research team at Tulane published a paper in the journal Science titled, "Synergistic Activation of Estrogen Receptor with Combinations of Environmental Chemicals" (Arnold et al. 1996).[6] The *San Francisco Chronicle* headlined its report of their findings, "With Pesticides, 1 Plus 1 Sometimes Equals 1,000." The results actually showed synergism multiplying effects of pesticide combinations up to 1600-fold over the substances tested individually. Lynn Goldman, EPA's director of the Office of Prevention, Pesticides and Toxic Substances commented on the study, "It is a very high priority for us to address the implications of this," indicating that the agency must consider how to test for effects of chemicals that might combine in the environment.

The study of synergistic interactions among substances is enormously complex because of the very high numbers of possible combinations. Even when research is narrowed to commonly associated groups—such as pesticides, herbicides, and defoliants used in agriculture and gardening or in the household—the complexity of risk assessment is still very high since many such chemicals accumulate in the soil, groundwater, and human or animal body tissue.

However, if researchers start with common combinations and classes of chemicals where they can postulate likely synergy, there is a basis for study. Relatively few studies on chemical synergism have been completed so far but the issue is clearly one that must be tracked by the public, by regulators, by the chemical industry, and by its customers (which really includes every industry).

Recent actions that indicate that the issue of synergism is likely to become much more visible. Richard Alexander, a worker and consumer protection attorney authored a paper on chemical synergism, citing studies done in relation to the Gulf War Syndrome (Alexander 1996). In March 1997, the Fifteenth National Pesticide Forum made synergistic interactions an organizing theme for a meeting of a coalition of activist groups. The combination of the

[6]In mid-1997 other scientists have not replicated these research results; however, other studies of synergistic interactions among toxins, including heavy metals, are emerging.

threat of litigation and activist initiatives should be enough to draw attention to this issue.

However complex they may be, the interrelated issues of endocrine disruption and synergism may become major challenges to risk assessment and drivers for fundamental rethinking of environmental regulation and management of our use of chemicals. Meeting these potential challenges calls for using industrial ecology concepts and methods to plan responses before the hypotheses involved are fully demonstrated. Such precautionary planning can manifest as a more systems-oriented approach to product and process design, which itself can enable a more competitive position for companies that embrace it.

A Broader Agenda for Reform

Industrial ecology methods bridge between short- and long-term solutions. Industrial ecologists propose a number of important criteria and measures of sustainability. They do not stop with the question, How can present products be made with greater efficiency of resource use? They also ask, *How can the life of products be extended? Are some products not needed? Can some companies move from selling products to delivering services? Do some technologies become sustainable through integration into more complete systems?*

Policies and regulations will need to evolve continually in order to facilitate rather than block innovative industrial and marketplace answers to such questions. *These are not necessarily long-term issues.* Walter Stahel points to a number of companies now applying his "advanced" ideas on product-life extension and the service economy. *See Chapter 4.*

European product take-back legislation is a first step toward motivating companies to adopt such concepts. While this legislation was still in the planning stage for their industry, auto manufacturers in France and Germany began responding. Several have created joint ventures for car disassembly and recycling. Their design teams are adding ease of disassembly and identification of materials into their criteria.

In an IE context, consideration of regulated product take-back legislation should begin with setting the broad objectives to be achieved. Ayres' industrial metabolism criteria and Stahel's product-life extension criteria could guide this process. With targets established, government policy makers could then initiate a voluntary process to design strategies with the relevant industries (such as electronics and appliances as well as automotive). *Challenge the manufacturers to come up with solutions even better than regulated product take-back.* Many companies have discovered that innovative early responses can forestall the need for regulated answers.

The Rhine Basin industrial metabolism study illustrates another way that a basic IE method can indicate directions for policy. This research on flows of heavy metals and selected toxic chemicals indicates that industry has made major reductions in its contributions to these flows. The major present sources are farms, consumers, runoff from roads and highways, and disposal sites. This suggests that education and incentives for farmers and consumers and research on how to decrease diffuse emissions from transportation are needed, not more regulation of industry. *See Chapter 3, Industrial Metabolism.*

"The new approach to environmental regulation . . . recognizes that attempts to micromanage a complex system from a single, centralized node are doomed to failure; dispersed control mechanisms and feedback loops are required."

(Graedel and Allenby 1995)

RESEARCH AND TECHNOLOGY POLICY

As with other areas of government activity, industrial ecology brings both useful perspectives to support technology policy making and a number of research needs to advance the field. Industrial ecology is one of six priority subjects of research identified by the National Research Council in a process involving an opinion survey, a public meeting, and discussions with experts. Other research issues on this list connect well with the broad definition of IE that we have used in this book; e.g., chemicals in the environment, energy, and environmental monitoring (National Research Council 1996).

Successful implementation of IE also requires a new level of integration across agencies and within them. In recent years there are a number of examples of the former such as the National Environmental Technology Strategy and the related creation of the Interagency Environmental Technology Office; the Global Change Research Program; Insight 2000 and the Rapid Commercialization Initiative. Within specific agencies integrative efforts include EPA's Common Sense Initiative and DOE's Industries of the Future program.

See discussion of organizational design requirements in Chapter 3.

IE Support of Technology Policy Development

Industrial ecology provides a framework for evaluating technical innovation in an integrated context of environmental and economic needs. With this perspective one can examine not just individual technologies but *the systems of technologies needed to achieve key environmental objectives in an economically feasible way*. The transportation cases at the end of Chapter 3, the Gambit Ltd. scenario at the end of Chapter 4, and the Developing Country Leapfrog scenario at the end of this book all illustrate this systemic approach to technological innovation.

Government support for technological development needs to be guided by awareness of the potential synergies among separate lines of inquiry. A breakthrough in one area may make possible commercial application of several other innovations.

Renewable energy is an outstanding example of a field where a whole systems view could dramatically lower greenhouse gas emissions and demand on non-renewable resources. Energy policy needs to track the timing of commercialization of new storage and transmission devices and a wide variety of renewable sources. Although not yet competitive in head-on competition with fossil sources, photovoltaics are now cost-effective in a number of niches, i.e. integrated into building components. They could play an effective role in a broader infrastructure design.

A decentralized energy infrastructure in a developing country could be cost-competitive by combining smaller, highly efficient fossil fuel plants, co-generation and energy cascading, wind, photovoltaics, passive solar, geothermal, and biomass sources. By avoiding the costs of building large centralized power plants and a new power grid, emerging technologies would be fully competitive.

Dynamic input-output modeling of this sort of systems strategy would enable policy makers and sources for development financing to assess its overall economic and environmental impact, compared with a conventional fossil fuel based solution.

Another example on a longer developmental path is economically viable fuel cell technology. This will depend on interrelated innovations and engineering improvements in the fuel cells themselves, reformer and membrane technologies, and storage and infrastructure systems for gas and/or hydrogen. R&D policy and funding needs to focus on the whole system required for this technology to be applied successfully.

To develop this systems approach to technology policy and research requires encouragement of interdisciplinary work in universities and Federal Labs, training in basic IE methods for both policy and research people, and educational outreach to financial institutions like the World Bank and venture capital funds.

See National Science and Technology Council 1994 and 1995 publications on technology for a sustainable world for detailed discussion of these and other themes.

IE Research Needs: Supporting Basic Research in Industrial Ecology

To support the transition to a sustainable mode of operation, we need to develop a much better understanding of the present industrial system and its impacts. The Environment, Health and Safety Committee of the Institute of Electrical and Electronics Engineers has identified key research needs (IEEB EHSC 1995). These include:

1. Engaging in studies to understand and model stocks, flows, and logistics of material movements throughout both the U.S. and global economies for all major industrial materials, including both renewables and nonrenewables. This includes mapping onto these flow models the environmental impacts and both human and ecosystem exposure data. These studies will support development of valid, efficient, risk-based environmental regulations and development of environmentally preferable products.
2. Developing industrial metabolism models of energy production and use, with overlays for assessing risks and evaluating alternative options for technical change. These models will support R&D policy and investment programs.
3. Developing integrated models of urban communities, including transportation, physical infrastructure, food, energy, and other systems. This would facilitate identification of major sources of environmental impacts, patterns of activities which give rise to them, and potential environmentally preferable technological or mitigation options.
4. Developing integrated models of industrial sectors of particular economic, environmental, or cultural importance, i.e., the agriculture, forestry, extractive, electronic and automotive sectors. This research would be important in supporting critical industries in reducing their environmental impacts without damaging their economic performance.

These four broad areas of research identified by the IEEE committee are complemented by research proposals focusing more on the ecological and social impacts of human activity and the sensitivity of ecosystems and global environmental systems. These are discussed extensively in *Industrial Ecology and Global Change* (Socolow et al. 1994).

Other important areas of research in support of industrial ecology are more process oriented. Methods and tools are needed for the evaluation of short-term options in a longer time frame; or the assessment of the implications of local action in a global context. A parallel need is research on the organizational implications of IE, including information systems design. In some cases all that will be needed is adapting existing approaches to the requirements of IE, much as TQM became a method for environmental quality or as design for x became the foundation for design for environment tools.

See Chapter 3 for more on IE as context for developing methods and tools and on organizational needs.

Another critical area of IE research will focus on the political and economic challenges of navigating the transition to a sustainable world. Our vision of clean industry operating in a cyclical, closed-loop mode implies significant changes in many sectors. How can policies be designed and implemented to support entrepreneurs and large corporations in discovering an economically viable path through the next decades of innovation?

PROCUREMENT

Government decisions have direct environmental impact in many domains, including the quality and specifications of government purchases, and the performance of governmental programs. Public agencies at every level can use IE to guide the purchasing of goods and services. The U.S. federal government has implemented a "buy-recycled" initiative across agencies. Several states have started programs to gain higher energy efficiency in government buildings.

Design of such specific actions could benefit from a systems overview of energy and materials flows through agencies and facilities. An eco-audit, using industrial metabolism criteria, could identify opportunities where changes in procurement would have the highest leverage

"The local government of Lower Saxony has a new environmental directive, intended to encourage the use of environmentally friendly products and to promote ecologically advantageous proposals from those seeking building contracts and contracts for other supplies and services. Where formerly economic criteria were applied exclusively in choosing among proposals, we have now added an ecological evaluation. It often turns out that when all social costs are taken into consideration, the most ecological solution is also the cheapest."

(Griefahn 1994)

for improvement of environmental performance. A foundation for this analysis would be setting performance objectives for total agency procurement using industrial metabolism measures of sustainability. *See Chapter 3.*

Government procurement practices can be used to help create markets for new integrated solutions, such as building energy systems. The City of Atlanta has declared itself a solar city and plans to help local solar energy manufacturers establish markets through long-term contracts for their products.

Or, a group of cities could contract for multi-year purchase of photovoltaic arrays for municipal facilities. The guaranteed market could enable manufacturers to increase manufacturing capability to a level needed for cost competitive pricing.

LOCAL GOVERNMENT

Architect William McDonough has designed a city/manufacture joint venture strategy to retrofit every window in a major city. The mix includes city industrial development bonds for initial financing; a mobile factory; job training program; manufacturing of energy efficient windows, and recycling of old windows into new glass and frames; cash rebates to residents, who then pay for the windows in a utility or tax bill; this provides millions of dollars in annual savings on utility bills; and, as the mobile factory moves on to the next city, the first city in the program retains an equity interest in the business.

In cities and counties are located mines, forests, industrial facilities, and farms. This is where we directly experience the impact of industrial activity on natural systems and where we also create our own impacts through consumption. Local governments themselves are responsible for public works that both manage and create many environmental impacts. US-EPA and DOE have developed initiatives to support local industrial ecology and sustainable development initiatives. The EPA's Urban and Economic Development Division web sites, listed under resources below, are useful beginning points for gathering information on what is available. DOE's Center for Environmental Excellence also maintains an excellent web site. Both have extensive information on industrial ecology and eco-industrial parks.

Many communities are now exploring the concept of sustainable development. Chattanooga has gone through a multi-year process of building a vision of itself as a sustainable community. Seattle is creating indicators and measures of local sustainable development.

Industrial ecology can support local leaders as they flesh out the concept of sustainable community. For example:

- Creating a community strategic plan for reducing the total waste stream and increasing efficiency of resource use (residential, commercial, public, and industrial)

◆ Extending the value of investments in the local infrastructure like landfills and water systems through more efficient use and recovery of resources

◆ Developing a highly effective industrial resource exchange providing markets for materials now discarded as wastes

◆ Strengthening economic development planning to encourage businesses that turn wasted resources into products and jobs

◆ Mobilizing educational and consulting resources to help the community's business and government operations increase energy efficiency and prevent pollution

◆ Streamlining environmental permitting and regulatory processes.

Communities following this path will gain many benefits: a cleaner environment, a stronger, more efficient economy, new jobs, and a reputation as a good site for starting a new business.

Analyzing the Industrial Metabolism of a Community

A city's economic development and public works agencies could work with university researchers to study the flows of materials and energy—the industrial metabolism of the area. Company and household surveys and analysis of flows into the landfill would be components.

The broad categories in a company survey might include the following:

◆ Products/services presently produced

◆ Process technologies used

◆ Material (water, feedstocks) and energy inputs

◆ Water, waste and energy outputs

◆ Cost of materials and disposal

◆ Environmental issues, including pollution prevention and energy efficiency initiatives.

The survey team would compile this information into an overview of company resources and needs in the region. Creating a regional knowledge base would be a first step

Big Plan to Cool Down Sacramento

Sacramento residents are planning to plant a million shade trees. Sacramento Municipal Utility District, in partnership with the nonprofit Sacramento Tree Foundation, has pledged to donate a half million shade trees to residents, schools and businesses by the year 2000. The energy saved will be enough to heat, cool, and light 4,000 homes for a year say officials at SMUD, which is spending $2 million a year to finance the program. The foundation's program will double the number of trees planted, making a total of 1 million trees. Nearly 100,000 trees in 5-gallon cans have been planted. The cost was about $40 per tree.

Trees in cities attack what scientists call the "urban summer heat island" where asphalt, cars, and breeze-blocking buildings actually increase temperatures up to eight degrees in the afternoon. Shade trees, by cooling streets and buildings and absorbing carbon dioxide, significantly reduce the effect. Studies by SMUD found that the cost benefit of energy conservation from tree planting should outweigh the utility's investment by about 2 to 1.

(*San Francisco Chronicle* 8-30-93)

toward a regional by-product exchange. A second phase of work could be setting up resources for each company to do a more comprehensive eco-audit.

A further audit would assess the potential for mining the community and regional "waste" streams to seek potential sources of by-product exchange and business development. Its components might include

- ◆ Percentages and tons of municipal landfill waste by type of material
- ◆ Characteristics of waste from specific sectors: residential, industrial, commercial, and institutional
- ◆ Materials listed by companies in local or state waste exchanges
- ◆ Types and amounts of materials collected in recycling programs
- ◆ Major sources of waste heat
- ◆ Sites and quantities of wastewater discharges
- ◆ Types of industries by SIC code
- ◆ Industrial input/output data.

Analysis of opportunities emerging from these data would point toward a broader economic development strategy, complementing recycling market development programs now in place. This would include setting community targets for reducing wastes and developing policies and programs that offer incentives and training to local business and citizens. The resulting economic development plan would emphasize attraction or generation of new businesses that could use major by-products available and those that turn former "wastes" into feedstocks for other companies or directly into products.

For example, recycling industrial park at the landfill itself could be a viable option. This could absorb major components of the community's waste stream.

Benefits for Municipal Infrastructure

Given the high costs of landfill development as a result of regulatory and technical requirements, significant cuts in waste streams can extend the life of this major community investment. Economic development agencies can make a case for using such savings to help fund other aspects of the project, as well as the development of recycling businesses.

In a similar way, return on water and wastewater infrastructure investments could be expanded by an industrial ecology strategy. Elements would be improving efficiency of use, pollution prevention programs in both industry and households, provision for cascading of water through different qualities of use, created wetlands for stormwater treatment, and ecological sewage treatment systems.

This approach applies to wastewater projects as well. A domestic wastewater treatment plant, for example, can maximize its investment through alternative technologies such as water efficient or waterless toilets, wastewater reuse and recycling, separation of stormwater from the waste stream, and using natural systems such as wetlands to treat certain wastewaters.

"If residue is generated . . . and then discarded, it violates one of the principles of good industrial ecology: that to the maximum extent possible, every molecule entering a manufacturing facility should leave that facility as part of a salable product . . . process residue streams, like products, should be designed to facilitate recovery and recycling."

(Graedel and Allenby 1995)

Another local initiative that flows naturally out of industrial ecology is restoration of ecosystems on public lands: infrastructure sites, parks, and the landscaping of public buildings. Participation in restoration projects, perhaps sponsored by corporations and civic clubs, would help preserve biodiversity while building awareness of ecological principles.

Green Plans

Norway, the Netherlands, Sweden, the U.K., and Canada have conducted comprehensive environmental and economic planning processes, generically known as Green Plans. The business communities in these countries have participated with government and non-governmental organizations in planning processes that have lead to benefits for both the environment and business. Regulatory structures have become more sophisticated, more effective, and less burdensome. Industrial environmental management has evolved to include more systemic pollution prevention measures and better patterns of resource utilization.

Green Plans set environmental aims and limits and provide performance demands to industry. Responsibility is placed on the industrial sectors to create the means of meeting these demands. They usually integrate economic incentives, green taxes, and tradable pollution permits as aspects of a comprehensive program.

Business has especially appreciated the greater certainty and predictability of future requirements. Having long-term but flexible strategic guidelines, managers are better able to plan investments and to discover innovative approaches.

A major conclusion of the green planning in the Netherlands is quite significant. Participants there emphasized that structural changes in patterns of consumption as well as production are needed. This connects with another central recommendation: develop channels to give precise feedback to people acting at the source of pollution or inappropriately using resources. Jan Hanhardt has pioneered the concept of Eco-Feedback in the Netherlands, including programs at the family and neighborhood levels.

See listing for the Resource Renewal Institute and Huey Johnson 1995 for more information on Green Plans.

RESOURCES ON GOVERNMENT'S ROLE

Organizations and Electronic Sites

Center for the Study of Environmental Endocrine Effects, 11 Dupont Circle, Washington, DC 20036. Telephone: 202-797-6368, Fax: 202-939-6969, Web: http://www.endocrine.org/95science. html.

Common Sense Initiative, US-EPA Office of Air and Radiation. Telephone: 202-260-2686, Fax: 202-260-2685.

DOE Center for Excellence in Sustainable Development: http://www.sustainable.doe.gov/industrial/index.html (includes links to many other sustainable community sites).

Global Vision, Community Metabolism page: http://www.igc.apc.global-vision.org/citymet.html.

IEEE paper on Sustainable Development and IE includes a large section on policy implications of IE: http://www.ieee.org/ehs/ehswp.html#SectI.

International Joint Commission: http://gopher.great-lakes.net:2200/1/partners/IJC.

President's Council on Sustainable Development: http://www.whitehouse.gov/WH/EOP/pcsd/index.html (see reports of sustainable communities and eco-efficiency task forces).

Resource Renewal Institute, Fort Mason Center, Building A, San Francisco, CA 94123. Telephone: 415-928-3774, Fax: -6529. For information on the Green Plan process, E-mail: info@rri.org, Web: http://www.rri.org.

Sustainable Community Sites: http://www.igc.apc.org/glencree/cityweb.html.

Sustainable Development Indicators: http://www.hq.nasa.gov/iwgsdi/.

University of Washington, Center for Sustainable Communities web links: http://weber.u.washington.edu/~common/hotlinks.html.

US-EPA homepage (http://www.epa.gov).

US-EPA Office of Sustainable Ecosystems and Communities. Telephone: 202-260-4002, Fax: 202-401-2213.

US-EPA Sustainable Communities Network: http://www.sustainable.org./SGN/sgn_index.html.

US-EPA Smartgrowth Network, Metropolitan Development that Serves Economy, Community, Environment: http://www.sustainable.org./SGN/sgn_index.html.

US-EPA Office of Urban and Economic Development.

U.S. Interagency Environmental Technologies Office, 955 L'Enfant Plaza North, SW, Ste. 5322, Washington, DC 20024. Internet: etstrategy@gnet.org.

Waitakere Online: http://nz.com/webnz/waitakere/ecocity/frameset.htm (a very extensive listing of sustainable city web sites world-wide).

Bibliographic Resources

Alexander, Richard, Esq. 1996. "Chemically Induced Diseases: Synergistic Effects and Cumulative Injuries Caused by Toxic Chemicals—Understanding the Gulf War Syndrome and Multiple Chemical Sensitivity [MCS]." Available from web site: http://seamless.com/talf/txt/article/gulfwar.html.

Allenby, Braden, and Cooper, William. 1994. "Understanding Industrial Ecology from a Biological Systems Perspective." *Total Quality Environmental Management*. Spring, pp. 343-354.

Anderson, Frederick R. 1994. "From Voluntary to Regulatory Pollution Prevention." In Allenby, Braden R., and Richards, Deanna J. (eds.). *The Greening of Industrial Ecosystems*. National Academy Press, Washington, DC, pp. 98-107.

Arnold, S.F., Klotz, D.M., Collins, B.M., Vonier, P.M., Guillette, L.J., Jr., and McLachlan, J.A. 1996. "Synergistic Activation of Estrogen Receptor with Combinations of Environmental Chemicals." *Science*, 272:1489-92. (Related news commentary: "Environmental Estrogens: New Yeast Study Finds Strength in Numbers," Kaiser, J., in same issue of *Science*, p. 1418.)

Colburn, Theo, Dumanoski, Dianne, and Myers, John Peterson. 1996. *Our Stolen Future*. Plume/Penguin Books. NY, London. (A survey of research on endocrine disrupters.)

Crisp, Thomas M. et al. 1997. *Special Report on Environmental Endocrine Disruption: An Effects Assessment and Analysis*. EPA/630/R-96/012. Washington, DC: Environmental Protection Agency, Risk Assessment Forum. Available via the Internet: http://www.epa.gov/ORD/webpubs/endocrine/.

de Graaf, John, and Harmann, Jack. 1995. Green Plans video. May be purchased or rented from The Video Project. Telephone:1-800-475-2638, Fax: 510-655-9115.

Global Environment and Technology Foundation will be releasing a CD-ROM on Technology for a Sustainable Future. Telephone: 703-750-6401.

Graedel, Thomas, et al. 1994. "Prioritizing Impacts in Industrial Ecology." In Socolow et al. *Industrial Ecology and Global Change*.

Graedel, T.E., and Allenby, B.R. 1995. *Industrial Ecology*. Englewood Cliffs, NJ: Prentice Hall, pp. 80-84 (for a discussion of policy and regulations).

Griefahn, Monika. 1994. "Initiatives in Lower Saxony to Link Ecology to Economy." In Socolow, Robert, et al., *Industrial Ecology and Global Change*. Cambridge University Press, NY, pp. 425-6.

Holland, John. 1996. *Hidden Order: Adaptation through Complexity*. Helix Books, Addison-Wesley, NY. (A very lucid general review of complex adaptive systems theory.)

Holmberg, John. 1995. *Socio-Ecological Principles and Indicators for Sustainability*. Institute of Physical Resource Theory, Chalmers University of Technology and Göteborg University, S-412 96 Göteborg, Sweden. (This publication contains the theoretical foundation for The Natural Step process. It may be ordered via the Internet: frt@fy.chalmers.se.)

Institute of Electrical and Electronics Engineers; Environment, Health and Safety Committee (IEEE EHSC). 1995. "White Paper on Sustainable Development and Industrial Ecology." (To obtain, contact: IEEE Technical Activities, 445 Hoes Lane, Piscataway, NJ 08855-1331. Telephone: 908-562-3908, Fax: 908-562-1769, E-mail:j.cerone@ieee.org.)

International Joint Commission. 1994. *Seventh Biennial Report on Great Lakes Water Quality* (Washington, DC, and Ottawa, Ontario) International Joint Commission, 1250 23rd Street, NW, Ste. 100, Washington, DC 20440. Telephone: 202-736-9000.

International Joint Commission. 1995. *Environmental Endocrine Effects: An Overview of the State of Scientific Knowledge and Uncertainties* (discussion draft first released for public comment at the September 22, 1995, public meeting of the Science Advisory Board of the U.S.-Canada IJC).

Johnson, Huey D. 1995. *Green Plans: Greenprint for Sustainability*. University of Nebraska Press, Lincoln, NE. Reports on implementation of Green Plan processes in Netherlands, Canada, New Zealand, and other countries. Telephone: 1-800-755-1105. Fax:1-800-526-2617 or from Resource Renewal Institute.

Lowe, Ernest A., et al. 1995. *Fieldbook for the Development of Eco-Industrial Parks*. Indigo Development. Volume 2, Final Report, Research Triangle Institute Project Number 6050, Research Triangle Park, NC. (To be released as US-EPA cooperative agreement project report.)

Lowe, Ernest A., Moran, Stephen R., and Holmes, Douglas B. 1997. *Eco-Industrial Parks: A Guidebook for Development Teams*. Indigo Development. (This is essentially the same text as the *Fieldbook for the Development of Eco-Industrial Parks* drafted by the same authors for US-EPA, with some updating and arranged in a sequence designed for EIP development teams. Available from Indigo. See the web site www.indigodev.com for ordering information.)

Martin, Sheila, et al. 1996. *Developing an Eco-Industrial Park: Supporting Research*. Volume 1, Final Report, Research Triangle Institute Project Number 6050, Research Triangle Park, NC. (This report includes "A Case Study of a Prototype EIP in Brownsville/Matamoros; Regulatory Issues and Approaches for Encouraging Eco-Industrial Park Development" and "Technologies Supporting Eco-Industrial Parks." To order, write RTI Information Services, POB 12194, Research Triangle Park, NC 27709.)

National Coalition Against the Misuse of Pesticides. 1997. *SYNERGY: Chemicals, Coalitions, and the Environment.* March 14-17, Washington, DC.

National Research Council, Policy Division. 1996. *Linking Science and Technology to Society's Environmental Goals*. National Academy Press. Washington, DC.

National Science and Technology Council. 1994. *Technology for a Sustainable Future* and *Bridge to a Sustainable Future*. National Environmental Technology Strategy. 1995. Telephone: 1-800-ENV-6676, E-mail: etstrategy@gnet.org.

Nelson, B.K., Conover, David L., and Lotz, W. Gregory. 1994. "Combined Chemical, Physical Hazards Make Exposure Harder to Calculate." *Occupational Safety and Health*, Vol. 63, No. 6, June, pp. 50-4.

New York Times. 1997. "Hormone Disruptors Require Additional Study, EPA Says." March 14, p. A26.

Pinchot, Gifford and Elizabeth. 1993. *The End of Bureaucracy and the Rise of the Intelligent Organization*. Berrett-Koehler Publishers, San Francisco, CA.

Porter, Michael E., and van der Linde, Claas. 1995A. "Green and Competitive: Ending the Stalemate." *Harvard Business Review*. September-October, pp. 122-134. Cambridge, MA. (A systems view of competitiveness with case studies and quantitative research indicating that well-designed environmental regulations drive innovation in the whole production process and product design. These innovations increase the productivity of resource use, building overall competitiveness. Includes criteria for regulations that will support this process.)

Porter, Michael, and van der Linde, Claas. 1995B. "Toward a New Conception of the Environment-Competitiveness Relationship." *Journal of Economic Perspectives ,9*, No. 4, Fall. (An extended discussion of the ways in which environmental regulations could change.)

Powers, Charles W., and Chertow, Marian R. 1997. "Industrial Ecology—Overcoming Policy Fragmentation." In Chertow, Marian R., and Esty, Daniel C., (eds.). *Thinking Ecologically: The Next Generation of Environmental Policy*. Yale University Press, New Haven, CT. (Papers generated from a series of four expert workshops held at Yale's Industrial Environmental Management Program in the Spring of 1996.)

Rolfe, Christopher J.B. 1994. *The Precautionary Principle, Statistical Power, and Improved Regulation*. Presentation to the British Columbia Water and Waste Association, January 19. (Available at http://opus.freenet.vancouver.bc.ca/local/wcel/wcelpub/6420.html.)

San Francisco Chronicle. 1996. "With Pesticides, 1 Plus 1 Sometimes Equals 1,000." June 7 (from Associated Press).

Socolow, R., Andrews, C., Berkhout, F., and Thomas, V. 1994. *Industrial Ecology and Global Change*. New York: Cambridge University Press. (For discussion of policy issues see pp. 359-370 and 401-468.)

Stigliani, W.M., and Anderberg, S. 1991. "Industrial Metabolism and the Rhine Basin." *Options*. International Institute for Applied Systems Analysis, Laxenberg, Austria, September.

Thomas, Margaret G. 1997. *Reinvention: Strategies for Sustainable Economic Development*. Midwest Research Institute, Kansas

City, MO. (This is a detailed sourcebook for local economic development and planning agencies. It includes options for action plans, bibliographies, and other resources in pollution prevention and waste reduction; recycling-based manufacturing and eco-industrial park development; energy efficiency investment; renewable energy development; and green business and environmental technology expansion. Order from Publications, MRI, 425 Volker Blvd., Kansas City, MO 64110-2299.)

US-EPA, Office of Solid Waste. 1994. *Review of Industrial Waste Exchanges*. Washington, DC. EPA-530-K-94-003. Telephone:1-800-424-9346 to order. For information on access to the National Materials Exchange Network, call 509-466-1532, Fax: 509-466-1041.

Weinberg, Matthew, et al. 1994. "Industrial Ecology: The Role of Government." In Allenby, Braden R., and Richards, Deanna J. (eds.). *The Greening of Industrial Ecosystems*. National Academy Press, Washington, DC, pp. 123-133.

Weitz, Keith, and Martin, Sheila, et al. 1995. "Regulatory Issues and Approaches for Encouraging Eco-Industrial Park Development." In *Developing an Eco-Industrial Park: Supporting Research*. Volume 1, Final Report, Research Triangle Institute Project Number 6050, Research Triangle Park, NC (to be released by US-EPA).

A FUTURE SCENARIO: Hometown Uses Industrial Ecology to Form its Economic Development Strategy

In October 1998, Marie Capistran returned from an economic development conference with a new idea: industrial ecology as a foundation for her community's economic planning. As Director of the Hometown Economic Development Council (HEDC), she was in a position to act on her new idea. She quickly briefed her Chamber of Commerce President, Hometown's Director of Public Works, a real estate developer, the CEO of the local utility, and her favorite anti-development activists.

She told each of them, "I think we can get past all the conflict we've had around industrial development in Hometown. This concept of industrial ecology tells me we can take care of our environment better than we have before while we grow strong new businesses."

"First, we get rid of the idea that we have to produce wastes. Our industry will be much stronger if we face facts:

plants are producing products they just haven't figured out how to sell. Then they pay good money to haul them to the dump! If they start seeing everything they produce as possible by-products, they'll be more competitive. Putting that into effect will give us a whole new type of business to recruit or incubate. We'll need firms that re-process the residual products."

Marie had soon recruited a leadership team to back her up in planning a community strategic planning conference: Making Hometown More Competitive through Industrial Ecology. In January 1999, industrial and union leaders, city officials, environmentalists, educators, and representatives of the state EPA met for two days. They focused on learning about their community as an industrial ecosystem and brainstorming initiatives.

Conference participants studied Kalundborg's industrial symbiosis and the early attempts to create eco-industrial parks. They mapped the industrial metabolism of Hometown, roughly charting the major energy and materials flows in the industrial, commercial, and residential sectors. (Participants had decided to stop using the word "waste" in Hometown by the end of the first day of the conference.)

Ecologists from the local university helped them develop a visual model of the major natural ecosystems in and around their city. In this model they highlighted areas of damage, vulnerability, and potential for restoration. Hometown Mayor, Andy Rice, said, "We have to know our environment's qualities, characteristics, and constraints if our economic development is going to work with it, not against it."

By the end of the conference, participants had committed to a process of creating a sustainable community strategic plan. They set initial broad objectives: reducing the total waste stream, increasing efficiency of energy use, lowering air and water emissions, and developing jobs. To carry this planning forward and assign specific targets they planned the following activities:

1. Continue the ecological modeling to deepen the understanding of Hometown's ecosystems and to guide setting environmental performance objectives
2. Survey companies to inventory their major unsold by-products and major inputs (energy as well as materials)

and to gather information on current implementation of pollution prevention and energy efficiency programs

3. Compile and review all existing data on materials going into the landfill
4. Create kits for family self-audits of their material and energy inputs and outputs
5. Refine the mapping of Hometown's industrial metabolism, using input from the above steps
6. Set up a community online information exchange with a web page, e-mail, discussion forums, and a data base of unsold by-products
7. Hold a second conference to review objectives and set specific targets within the broad environmental and economic objectives set in the first meeting.

Hometown's TV and radio stations gave the conference full coverage and committed to reporting regularly on each of the initiatives. The *Hometown Times* asked Marie to write a weekly column on the project and KHOM scheduled a regular call-in show to keep channels open for citizen input.

Marie Capistran's board was willing to invest seed money and staff time into these projects. She argued that a cleaner, more efficient industrial sector would create new business development opportunities, make Hometown a more attractive site for plant location teams, and strengthen the town's economy.

The local banks and utilities also contributed funds and staff time for number crunching. A state foundation made grants to the university, community college, and K-12 system to support the survey work and analysis. Finally, the City Council voted general fund money for project coordination and outside consultants.

The State EPA and Energy Commission helped the schools design energy efficiency and pollution prevention training programs.

The schools at every level played a very important role in the data collection and processing. Student/faculty teams, augmented by engineering consultants, worked together in the company surveys to determine the major industrial by-products available for recycling or reuse. They used a geographic information system to map the major opportunities and possible customers.

A business school team created a business plan for a resource exchange operating on the principles of investment recovery. High school and elementary school children made sure their families did the household audits and compiled the data they gathered from them.

By late June 1997, Hometown had enough information to hold its second conference. The surveys had uncovered significant opportunities for increasing the efficiency of energy and materials use in industrial plants and for turning wastes into useful products. The ecological modeling highlighted natural constraints to guide the setting of specific targets for environmental performance and development.

Following the second conference, the Economic Development Council used the Public Works waste stream and landfill analysis to identify six early business development opportunities in materials and chemical re-processing. Working with the university and community college, HEDC created a business incubator.

The utility recruited a food processing plant, a greenhouse, and an aquaculture firm to locate on its land to use excess steam from its power plant. An investment recovery company decided to open an office to manage the exchange of energy and material by-products.

In the fall of 1997, a local real estate developer negotiated a deal with the city to create an eco-industrial park on closed landfill property. The analysis of community waste streams gave her the foundation for a recruitment strategy based on low-cost supply streams.

Engineers at the university helped find technologies for turning several major waste flows into usable by-products. HEDC supported the creation of industrial development bonds by the city. The development of the Hometown Ecopark moved forward as a strong public/private partnership.

By 2002, city leaders from around the world were coming to Hometown to learn how industrial ecology could help make local economic development sustainable. The economy was strong. Eighty new local businesses interacted in the city's industrial ecosystem with 12 plants of major corporations. All operated with environmental performance standards higher than those required by regulations.

7 Industrial Ecosystems and Eco-Industrial Parks

INTRODUCTION

The concepts of industrial ecosystems and eco-industrial parks (EIPs) have embodied the industrial ecology approach in very concrete terms. The story of Kalundborg, an industrial ecosystem identified in Denmark, has circled the planet as a sign that industry can coexist with nature in a more benign manner.[1] Other industrial regions are attempting to emulate this pattern of by-product exchange.

Deanna Richards at the National Academy of Engineering titled the second book on IE, *The Greening of Industrial Ecosystems*. This suggests that there is merit in viewing all industrial clusters as ecosystems (though they may often be far from benign in their performance).

Communities in the U.S. have adopted the allied concept of eco-industrial parks, and the President's Council on Sustainable Development has named four of them demonstration sites for EIP development. Three years after this concept was introduced, at least 15 communities in North America have started industrial ecosystem or EIP projects.

". . . the traditional model of industrial activity—in which individual manufacturing processes take in raw materials and generate products to be sold plus waste to be disposed of—should be transformed into a more integrated model: an industrial ecosystem.

"In such a system the consumption of energy and materials is optimized, waste generation is minimized and the effluents of one process . . . serve as the raw material for another process."

(Frosch and Gallopolous 1989)

[1]The plant managers at Kalundborg are the first to admit that they have made only a modest beginning on improving environmental performance.

INDUSTRIAL ECOSYSTEMS

The proposal that the design of industrial systems may benefit from understanding natural system dynamics has suggested the concept of industrial ecosystems. Nick Gertler gives a specific definition in his dissertation on the conditions for developing this sort of network.

> An industrial ecosystem is a community or network of companies and other organizations in a region who chose to interact by exchanging and making use of by-products and/or energy in a way that provides one or more of the following benefits over traditional, non-linked operations:
>
> - Reduction in the use of virgin materials as resource inputs
> - Reduction in pollution
> - Increased systemic energy efficiency leading to reduced systemic energy use
> - Reduction in the volume of waste products requiring disposal (with the added benefit of preventing disposal-related pollution)
> - Increase in the amount and types of process outputs that have market value. (Gertler 1995, Chapter 1)

Gertler makes a valuable distinction between two basic IE strategies for moving toward a closed-loop or cyclical industrial system. One focuses on products themselves through product policy, life-cycle assessment, design for environment, and product-life extension. Action in this mode is independent of location and tends to focus within a company or industry. The second IE strategy seeks to optimize materials and energy flows among facilities within specific regions or industrial ecosystems. The focus is on the inputs and outputs of a set of production systems in different companies and agencies.

The two approaches are largely complementary ways of increasing resource efficiency and reducing pollution. However, at times managers and designers will need to decide on trade-offs in costs and benefits (environmental and economic) between them. For example, a power plant staff may want to consider fuel input or process changes that reduce or eliminate by-products the plant sells to other companies (gypsum and fly-ash). The trading of wastes as by-products is not a good in itself if there are more effective solutions upstream.

In an industrial ecosystem, the interactions among companies resemble the dynamics of natural ecosystems, where all materials are continually reused and recycled. Major petrochemical complexes are classic examples of industrial ecosystems operating within an industrial sector. The chemical industry has always sought uses and customers for unmarketed products. This practice has increased efficiency of resource use but has not necessarily been related to conscious elimination of pollution until recent decades. *See the sidebar on the Houston Ship Channel.*

The Industrial Symbiosis at Kalundborg

One of the favorite cases presented by industrial ecologists is the story of the spontaneous but slow evolution of the "industrial symbiosis" at Kalundborg, Denmark.[2] This web of materials and energy exchanges among companies (and with the community) has developed over the last 20 years in a small industrial zone on the coast, 75 miles west of Copenhagen. Originally, the motivation behind most of the exchanges was to reduce costs by seeking income-producing uses for "waste" products. Gradually, the managers and town residents realized they were generating environmental benefits as well, through their transactions.

The Kalundborg system comprises five core partners:

- Asnaes Power Station—Denmark's largest power station, coal-fired, 1,500 megawatts capacity
- Statoil Refinery—Denmark's largest, with a capacity of 3.2 million tons/yr (increasing to 4.8 million tons/yr)
- Gyproc—a plasterboard factory, making 14 million square meters of gypsum wallboard annually [roughly enough to build all the houses in six towns the size of Kalundborg]

[2]"Industrial symbiosis: A cooperation between different industries by which the presence of each increases the viability/profitability of the other(s), and by which the demands of society for resource savings and environmental protection are considered. Symbiosis is the living together of dissimilar organisms in any of various mutually beneficial relationships. Here the term is used to mean industrial cooperation with mutual utilization of residual products." From *Industrial Symbiosis*, a publication of the Kalundborg companies. No date.

The Houston Ship Channel

The Houston Ship Channel is a large petrochemical complex where companies have long turned their wastes into raw materials for their own or other companies' processes.

The Channel stretches dozens of miles, making this inland city one of the largest harbors in the United States. Numerous chemical, petrochemical, and energy production facilities are located on each side of the ship channel.

These industries exchange products that are generated as by-products of their primary production activities. Energy companies purify raw natural gas, for example, to remove constituents (carbon dioxide, hydrogen, nitrogen, ethane, propane, and ethylene) that are then either used on site, or sent by pipeline for use in neighboring plant processes.

The marketplace for exchanges of "by-products" (versus what might be "wastes" in another industrial system) is active among all companies in the Houston Ship Channel complex. Economic, not environmental, forces encouraged companies to strive for these efficient operations.

(Douglas Holmes 1993, personal communication)

- Novo Nordisk—an international biotechnological company, with annual sales over $2 billion. The plant at Kalundborg is their largest, and produces pharmaceuticals (including 40% of the world's supply of insulin) and industrial enzymes
- The City of Kalundborg—a city that supplies district heating to the 20,000 residents, as well as water to the homes and industries.

Over the last two decades, these partners spontaneously developed a series of *bilateral exchanges* which also include a number of other companies. There was no initial planning of the overall network; it just evolved as a collection of one-to-one deals that made economic sense for the pairs of participants in each.

Energy Flows

The Asnaes power station is coal fired and operates at about 40 percent thermal efficiency. Like all other fossil-fuel power stations, the majority of energy generated still goes up the stack. Another large energy user, the Statoil refinery flared off most of its gas by-product. Then, starting in the early 70s, a series of deals were struck:

- The refinery agreed to provide excess gas to Gyproc, whose management had recognized that the burning gas from Statoil's flares was a potential low-cost fuel source.
- Asnaes began to supply the city with steam for its new district heating system in 1981 and then added Novo Nordisk and Statoil as customers for steam. The district heating, encouraged by the city and Danish government, replaced about 3,500 oil furnaces (a significant non-point source of air pollution).
- The power plant started using salt water, from the fjord, for some of its cooling needs. By doing so, it reduced the withdrawals of fresh water from Lake Tissø. The resulting by-product is hot salt water, a small portion of which is supplied to the fish farm's 57 ponds.
- In 1992, the power plant began substituting fuels, using surplus refinery gas in place of some coal. This only became possible after Statoil built a sulfur recovery unit to comply with regulations on sulfur emission; the gas was then clean enough to permit use at the power plant.

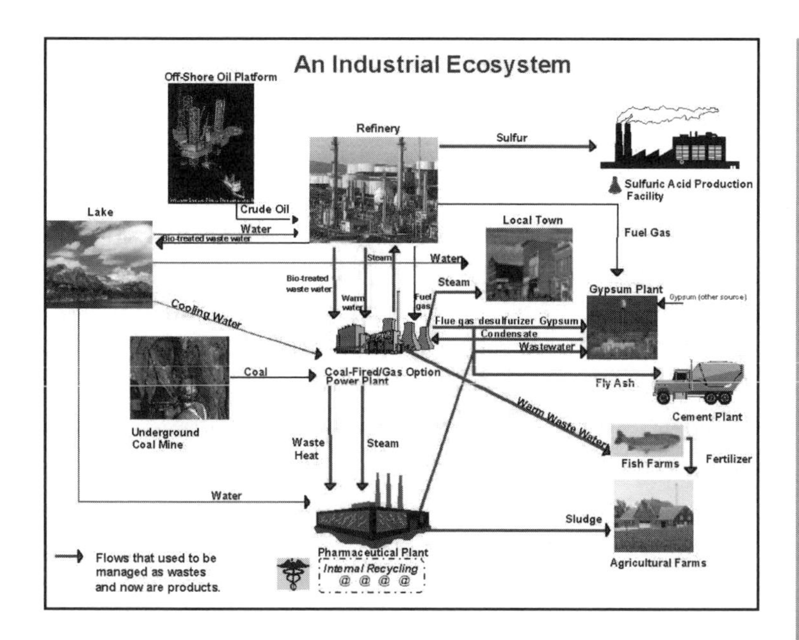

Materials Flows

In 1976 the Novo-Nordisk plant started the pattern of materials flows, matching the evolving energy flows at Kalundborg.

- ◆ Sludge from Novo Nordisk's processes and from the fish farm's water treatment plant was offered as fertilizer to nearby farms. This is now a large portion of the entire Kalundborg exchange network, totaling over 1 million tons per year.
- ◆ A cement company uses the power plant's desulfurized fly ash.

Asnaes reacts the SO_2 in its stack gas with calcium carbonate, thereby making calcium sulfate (gypsum), which it sells to Gyproc, supplying two-thirds of the latter's needs.

- ◆ The refinery's desulfurization operation produces pure liquid sulfur, which is trucked to Kemira, a sulfuric acid producer.
- ◆ Surplus yeast from insulin production at Novo Nordisk goes to farmers as pig food.

This web of recycling and reuse has generated new revenues and cost savings for the companies involved and reduced pollution to air, water, and land in the region. In ecological terms, Kalundborg exhibits the characteristics of a simple food web: organisms consume each other's waste materials and energy, thereby becoming interdependent with each other.

This pattern of intercompany reuse and recycling has reduced air, water, and ground pollution, conserved water and other resources, and generated new revenue streams from the by-products exchanged.

Through 1993, the $60 million investment in infrastructure (to transport energy and materials) has produced $120 million in revenues and cost-savings.

Lessons From Kalundborg

What can we learn from the Danes' experience over the past two decades? Here are some responses from the managers directly involved:

- All contracts have been negotiated on a bilateral basis.
- Each contract has resulted from the conclusion by both companies involved that the project would be economically attractive.
- Opportunities not within a company's core business, no matter how environmentally attractive, have not been acted upon.
- Each partner does its best to ensure that risks are minimized.
- Each company evaluates its own deals independently; there is no system-wide evaluation of performance, and they all seem to feel this would be difficult to achieve.

Jørgen Christensen, former Vice President of Novo Nordisk at Kalundborg, identifies several conditions that are desirable for a similar web of exchanges to develop:

- Industries must be different and yet must fit each other.
- Arrangements must be commercially sound and profitable.
- Development must be voluntary, in close collaboration with regulatory agencies.

- A short physical distance between the partners is necessary for economy of transportation (with heat and some materials).
- At Kalundborg, the managers at different plants all know each other.

See Gertler 1995 or Lowe 1997, Appendix 1, for a more detailed description of the exchanges and the conditions that supported development of the pattern.

A Recycling Network in Styria

For several years no researchers identified comparable industrial ecosystems. However, Erich Schwarz[3] has discovered and studied a much larger, more diverse "industrial recycling network" in the Austrian province of Styria (Schwarz, Erich, and Steininger, Karl W. 1995; and Schwarz 1994). The research started by tracing the by-product inputs and outputs of two major enterprises and soon found a complex network of exchanges among over 50 facilities. Industries participating include agriculture, food processing, plastics, fabrics, paper, energy, metal processing, wood working, building materials, and a variety of waste processors and dealers.

Materials traded in the Styrian network include the familiar recyclables like paper, power plant gypsum, iron scrap, used oil, and tires, as well as a wide range of other by-products. Schwarz does not provide details on activities relating to energy cogeneration or cascading, but they do play a role in the network.

The plant managers in Styria were not aware of the larger pattern of exchange that has evolved. They were motivated purely by the revenues from by-products they could sell and the savings in landfill disposal costs for either sold or free outputs. In some cases the by-products were less expensive or higher quality than primary materials would be.

The Styrian recycling network suggests that Kalundborg may be unique only in the level of awareness developed there. There may be many other spontaneously occurring

[3]Dr. Schwarz is a researcher at Institute für Innovationsmanagement, Karl-Franzens University, Graz.

industrial ecosystems with significant flows of materials and energy among different companies. However they can become more effective by calling attention to the patterns of trade and making information on resources and needs available. We will discuss further means for doing this.

Industrial Ecosystem Precursors

Beginning in the 70s, several researchers developed conceptual models of local by-products exchanges, out of the search for greater resource efficiency generated by the oil crises. Industrial metabolism pioneer, Robert Ayres, describes Korean, Polish, and U.S. proposals in a recent paper (Ayres 1996). For instance, the Korean project concept linked a coal-fired power plant, an aluminum smelting operation, and a concrete kiln.

Nelson Nemerow proposed creating complexes of co-located companies using each other's materials and energy "wastes" as feedstocks (also beginning in the 70s). His phrase for this is "Environmentally Balanced Industrial Complex (EBIC)." In his recent book (Nemerow 1995) he described 10 possible EBICs involving 16 different industries. Nemerow analyzes the way these potential complexes of plants can become mini-"foodwebs," reducing pollution/waste generation. For instance, four of his complexes focus on connecting fossil fuel power plants to consumers of waste heat and material by-products. In these complexes the partners include cement and cement block, lime, and agricultural facilities. (In defining zero pollution, Nemerow does not account for the CO_2 emissions or the residual emissions from the scrubbers and other treatment technologies.)

A research team under the direction of Jack Spurlock at Georgia Tech evolved a third industrial ecosystem model in response to the first energy crisis. Again, the focus was on identifying strategies to promote the flow of energy and materials among plants co-sited with power plants (Spurlock 1980).

Greening Industrial Ecosystems

Exploration of the evolution of the industrial ecosystems at Kalundborg and elsewhere has led industrial ecologists to ask, what conditions are needed to optimize exchange of

The Zero Emissions Research Initiative

The Zero Emissions Research Initiative (ZERI) has targeted zero waste as a means of improving the profitability of manufacturing while reducing its environmental impact. Like industrial ecologists, ZERI envisions multi-industry clusters of factories in which emissions from one plant are used as inputs by another. The initiative's director, Gunter Pauli, was inspired by the quality movement's zero defects goal. ZERI seeks to achieve zero emissions, at least in terms of releases at a particular site. Eco-industrial park projects at Chattanooga and Cape Charles have incorporated this goal, as have other industrial development projects in Asia and South Africa.

Participants have outlined a detailed research agenda and demonstration projects to achieve the technical breakthroughs needed to achieve the ambitious goal. ZERI, based in the United Nations University's Tokyo headquarters, coordinates the work of a global network of researchers and industrialists. ZERI Foundation memberships by companies fund the R&D. In turn, the companies will benefit from priority access to research results and by participation in ZERI projects.

Initial ZERI projects focus on industries generating large biomass wastestreams. These include the recovery of high-value green biomass from logging residue; the redesign of paper recycling for zero emissions; and the clustering of beer brewing and fish farming as complementary industries for the elimination of waste.

(Pauli 1995)

resources among a network or cluster of companies. The broad outlines of a response are emerging in the work of several researchers (Ayres 1996, Bass 1997, Côté 1994, Cohen-Rosenthal 1995, Gertler 1995, Lowe 1995, and Schwarz 1994 and 1995).

The Three Conditions

Understanding of the industrial ecosystem concept: While pairs of companies will continue to seek exchanges without the concept, the greatest benefits will come from a more systemic approach. Workshops can alert managers to the potential benefits, report on cases demonstrating the economic and environmental values of resource exchange, and set objectives.

Existence of an organizing structure: A high performance industrial ecosystem needs some agent concerned with the network as a whole. Individual companies will self-organize to manage their own exchanges. An organizing office could support them while seeking opportunities for optimizing the larger system. *See The Organizing Structure below.*

Screening: Prospects for the network will want to know they can trust the other companies to perform capably as suppliers or customers, as well as in all matters of environmental management relating to resource flows between them.

Assurance of future support: Companies who recognize the value to them of being in a by-product exchange network will want to know that there is a structure committed to maintaining the system. Future recruitment to fill vacancies, assist firms that have lost a by-product supplier or customer, and coordinate exchanges outside the region are some ways to meet this need.

Flexibility: The network will grow, participants will change, and environmental objectives will need to be adjusted in light of new research information or regulations. For an industrial ecosystem to remain viable its members and organizing office will need to remain flexible and adaptive, learning from errors as well as successes.

The Organizing Structure

An organization responsible for *facilitating* operation of the whole system will be necessary to achieve the highest economic and environmental benefits in an industrial ecosystem. This entity would seek to optimize performance through information flows, research, brokering relationships, negotiating with regulators, and encouraging new business development. It could be either a public or private organization.

A public entity could be formed by the three areas of government whose interests intersect in the industrial ecosystem concept: economic development, public works, and environmental protection. Optimal performance of a resource exchange network improves overall business performance and opens new economic development opportunities. Major reductions in waste streams reduce demands on municipal infrastructure such as landfill and water or sewage systems (major community investments). All of these changes reduce environmental burdens. An organizing office for an industrial ecosystem could be formed by collaboration between agencies in these three domains.

A good private sector alternative would be an investment recovery business. Practitioners in this new field seek highest value reuse of all surplus resources in a system. This would be a more effective approach than simply relying on waste brokers, who focus on the higher value materials. (See Chapter 5, Business Opportunities, for description of this field.) If the industrial ecosystem were centered in an industrial park, the company managing that property could assume this organizing role for the park and surrounding region.

Industrial waste or resource exchanges are now operating at regional and national levels. They would provide valuable backup to more local exchange networks.

A Resource Survey

Whether public or private, the organizing office would survey resources and conduct educational workshops at the beginning of forming an industrial ecosystem.

The steps in the survey would include:

1. Analyze the material and energy inputs and outputs of major industries in the area, such as:
   - Composition and nature of flows of materials and energy
   - Amounts

- Distribution of flows in time (steady, periodic, episodic, or irregular)
- Material and energy quality (consistency over time and purity).

2. Assess the business potential of pooling small quantities of some materials to create flows sufficient to market.
3. Disseminate information and check for matches with existing businesses.
4. Determine material or energy processing required to achieve quality requirements. What new businesses may be needed for reprocessing and handling?
5. Identify potential customer industries (as candidates for recruitment or incubation) to utilize existing material or energy flows.
6. Define volume requirements of potential customers.
7. Make this information available to all firms in the region (and on regional exchange data bases).
8. On the basis of the results, adjust the network strategy.

An industrial ecosystem project team could sponsor workshops where firms can explore potential exchanges, while learning about other industrial ecology strategies. This will also allow them to explore other possible business and environmental benefits they could gain from collaboration.

A valuable tool for the organizing team and participating companies would be an evolving computer model of the exchange network, supported by materials and energy data bases.[4] This would enable them to simulate process interactions within the network of companies, step by step. The survey of community industries would define initial input needs and by-products that other companies might consume. As other companies commit, the list of potential exchanges would grow.

The model would allow each company to test the feasibility of possible trades and to simulate potential impacts of technical and business changes on their exchanges. The

[4] A pioneering study in the late 1970s by researchers at Georgia Tech (Spurlock and Ward 1980) defines components of a model for "synergistic co-siting of industrial facilities."

organizing team would be able to use it to *work toward* optimization of the whole system.

A very high level of optimization in a complex system is itself complex and not necessarily achievable. The goal here is to seek better results than those possible if the system relied solely on spontaneous bilateral deals between firms.

The success of the network will largely depend upon the self-organizing interactions among participants. The organizing office cannot control these interactions, but it can offer the *conditions* that will help companies discover the value of working together and seeking collective benefits beyond what they can achieve in isolation.

A Broader View of Industrial Ecosystems

Many enthusiasts have seized upon one element of ecosystem dynamics from the Frosch and Gallopolous definition, *"the effluents of one process . . . serve as the raw material for another process."* As a result, the concept of industrial ecosystems and the field of IE itself are often reduced to systems for trading wastes as by-products. In many cases, optimal solutions will call for eliminating wastes earlier in the cycle rather than trading them.

Ecosystems in nature demonstrate many other strategies relevant to industry and to designers. *See Chapter 2 on Living Systems for discussion.* These include:

- The sole source of power for ecosystems is solar energy.
- In natural systems there is no such thing as "waste" in the sense of something that cannot be absorbed constructively somewhere else in the system. Nutrients for one species are derived from the death and decay of another.
- A major portion of energy flows in ecosystems is consumed in decomposition processes that make nutrients in wastes available.
- Concentrated toxic materials are generated and used locally.
- Efficiency and productivity are in dynamic balance with resiliency. This balance preserves the ability of ecosystems to adapt and evolve.

- Ecosystems remain resilient in the face of change through high biodiversity of species, organized in complex webs of relationships.
- The many relationships are maintained through self-organizing processes, not top-down control.
- In an ecosystem, each individual in a species acts independently, yet its activity patterns cooperatively mesh with the patterns of other species. *Cooperation and competition are interlinked and held in balance.*

As an industrial ecosystem evolves, members can draw upon these and other ecological dynamics to guide their continuous improvement. They are potentially useful in "designing" interactions between public facilities and service companies as well as manufacturing plants. They are also very congruent with the growing partnership and networking relationships among companies, large and small.

Challenges and Risks in Creating By-Product Exchanges

Companies using each other's residual products as inputs face the risk of losing a critical supply or market if a plant closes down or changes its product mix.

To some extent, this risk can be managed as with any supplier or customer relationship (i.e., keep in touch with alternative suppliers and write contracts ensuring reliability of supply and including recourse if obligations are not met).

Proprietary information could become available to competitors.

Information about a company's inputs and outputs can be used to understand and copy proprietary production processes. Waste exchanges usually do not name the source of materials until a buyer wants to connect with them; but in a local area, it would be easier to identify the source of materials in exchange.

Uneven quality of by-product materials could cause damage to equipment or poor products.

Handling this issue is a fairly standard contracting procedure for any supplier relationship. Dealing with a supplier in an industrial ecosystem would be no different. Both park management and tenant companies would need strong quality control standards and processes.

Exchange of by-products could lock in continued reliance on toxic materials.

The pollution prevention solutions of materials substitution or process redesign should take priority over trading toxics within a resource network. Companies should have continuing support from environmental agencies or consultants in finding ways to eliminate toxics from their operations.

Possible innovations in regulation to enable industrial ecosystem development may not be allowed by regulatory agencies.

The regulatory changes needed to support industrial ecosystem development are in line with current trends in regulatory policy. Perhaps umbrella permitting could be designed for different groups, so that companies with very different levels of risk would be under different permits.

ECO-INDUSTRIAL PARKS (EIP)

A specific form of industrial ecosystem is one focused in an existing or new industrial park. This focus opens additional opportunities for improving economic and environmental performance. A US-EPA research project defined this as follows:

> An eco-industrial park is a community of manufacturing and service businesses seeking enhanced environmental and economic performance through collaboration in managing environmental and resource issues including energy, water, and materials. By working together, the community of businesses seeks a collective benefit that is greater than the sum of the individual benefits each company would realize if it optimized its individual performance only.
>
> The goal of an EIP is to improve the economic performance of the participating companies while minimizing their environmental impact. Components of this approach include new or retrofitted design of park infrastructure and plants; pollution prevention; energy efficiency; and intercompany partnering. Through collaboration, this community of companies becomes an "industrial ecosystem" (Lowe et al. 1997A).

Some developers and communities have used the term eco-industrial park in a relatively loose fashion. We encourage applying this term to developments that are more than

- ◆ a single by-product exchange pattern or network of exchanges
- ◆ a recycling business cluster (resource recovery, or recycling companies)
- ◆ a collection of environmental technology companies
- ◆ a collection of companies making "green" products
- ◆ an industrial park designed around a single environmental theme (i.e., a solar energy driven park)
- ◆ a park with environmentally friendly infrastructure or construction
- ◆ a mixed-use development (industrial, commercial, and residential).

An EIP may *include* any of these features. But the critical element in defining an EIP is the interactions among its member businesses and between them and their natural environment.

The EIP—A Menu of Opportunities

Some EIP designers have tended to emphasize one strategy over all others—the exchange of by-products between companies in an industrial park. Although this "closing the loop" approach is an important factor, it is but one of a number of elements in eco-industrial park design.

There is a rich menu of design options, including site design, park infrastructure, individual facilities, and shared support services. Some major strategies an EIP design team can draw upon in planning a park are highlighted on page 144.

Construction/Rehabilitation

New construction or rehabilitation of existing buildings follows the best environmental practices in materials selection and building technology. These include recycling or reuse of materials and consideration of life cycle environmental implications of materials and technologies.

Strategies for Designing an Eco-Industrial Park

Several basic strategies are fundamental to developing an EIP. Individually, each adds value; together they form a whole greater than the sum of its parts.

Integration into Natural Systems

- Design the EIP in harmony with the characteristics and constraints of local ecosystems
- Design landscaping in terms of restoration of native ecosystems where feasible
- Minimize contributions to global environmental impacts, e.g., greenhouse gas emissions.

Energy Systems

Maximize energy efficiency through facility design or rehabilitation, cogeneration,[5] energy cascading,[6] and other means as well as

- Achieve higher efficiency through interplant energy flows
- Use renewable sources whenever feasible.

Materials Flows and "Waste" Management for the Whole Site

- Emphasize pollution prevention, especially with toxics
- Ensure maximum reuse and recycling of materials among EIP businesses
- Reduce toxic materials risks through integrated site-level waste treatment
- Link the EIP to companies in the surrounding region as consumers and generators of usable by-products via resource exchanges and recycling networks.

Water

Design water flows to conserve resources and reduce pollution through strategies similar to those described for energy and materials.

Effective EIP Management

In addition to standard park service, recruitment, and maintenance functions, EIP management does the following:

- Maintains the mix of companies needed to best use each other's by-products as companies change
- Supports improvement in environmental performance for individual companies and the park as a whole
- Operates a site-wide information system that supports intercompany communications, informs members of local environmental conditions, and provides feedback on EIP performance
- Offers value-adding services and facilities, such as environmental management, a dining and meeting commons, daycare, or procurement of common supplies.

---

[5]Cogeneration is the capturing and using of otherwise "wasted" heat from the electrical generating process.

[6]Energy cascading is using residual heat in liquids or steam from a primary process to provide heating or cooling to a later process. For example, steam from a power plant is used in a district heating system.

Natural Systems—An industrial park can fit into its natural setting in a way that minimizes environmental impacts while cutting some costs of operation. The Herman Miller Phoenix Design plant illustrates the use of native plant reforestation and the creation of wetlands to minimize landscape maintenance, purify stormwater runoff, and provide climate protection for the building. These and other natural design concepts can be used throughout an industrial park.

At another level, design choices in materials, infrastructure, and building equipment, plant design, and landscaping can reduce a park's contributions to global climate change and consumption of nonrenewable resources.

Energy—More efficient use of energy is a major strategy for cutting costs and reducing burdens on the environment. In EIPs, companies seek greater efficiency in individual building, lighting, and equipment design; flows of steam or heated water from one plant to another (energy cascading); and possibly connections into district heating of homes. (In power plants and many industrial processes, the majority of heat generated goes up the stack or to a cooling tower rather than producing value.) In many regions, the park infrastructure can use renewable energy sources such as wind and solar energy.

Materials Flows—In an eco-park, companies perceive wastes as products they have not figured out how to reuse internally or market to someone else. Individually, and as a community, they work to optimize use of all materials and to minimize the use of toxic materials. The park infrastructure may include the means for moving by-products from one plant to another; warehousing by-products for shipment to external customers; and common toxic waste processing facilities.

Water Flows—In individual plants, designers specify high efficiency building and process equipment. Process water from one plant may be reused by another (water cascading), passing through a pretreatment plant as needed. The park infrastructure may include mains for several grades of water (depending on the needs of the companies) and provisions for collecting and using stormwater run off.

Park Management and Support Services—As a community of companies, an EIP needs a more sophisticated management and support system than a traditional industrial park. Management supports the exchange of by-products among companies and helps them adapt to changes in the mix of companies (such as a supplier or customer moving out) through its recruitment responsibility. It may maintain links into regional by-product exchanges and a site-wide telecommunications system. The park may include shared support services such as a training center, cafeteria, day-care center, office for purchasing common supplies, or transportation logistics office. Companies can add to their savings by sharing the costs of these benefits.

Sustainable Design and Construction—EIP planners design buildings and infrastructure to optimize the efficient use of resources and to minimize pollution generation. They seek to minimize the impact on the ecosystem by careful site preparation and building plants and park systems. The whole park is designed to be durable, maintainable, and readily reconfigured to adapt to change. At the end of its life, materials and systems can be easily reused or recycled.

BUILDING THE COMMUNITY OF COMPANIES

We believe that tenants in an EIP will benefit from collaboration—from working together as a community. There are many signs that large, medium, and small firms are functioning in closer partnerships as a way of building competitive advantage. Managers at Kalundborg have said that their close sense of community was essential to the development of their by-product exchange network.

The developer and manager (if different) of an eco-park only need to build upon an existing trend to support tenants in forming their community. The key method is quite simple: self-organization. With the right context for employees from different companies to get together, they will figure out how they can benefit from working together. Park management can provide events and tools to support the process. These could include:

- Conferences for possible recruits and actual future tenants
- A computer bulletin board where future tenants can start getting acquainted and explore beneficial ways of working together (This communication then links into the EIP information system once they have moved in.)

- A welcoming party (and informal idea session) as each new tenant moves in
- Creation of a tenants association as a community system of governance (including conflict resolution capabilities)
- Education in the flexible network (value adding network) concept for business collaboration.

If the team has succeeded in building a significant by-product exchange network among the tenants, their process of negotiating contracts and implementing the exchanges will contribute to the sense of community. (Park management may play a role in supporting the ongoing viability of the network and identifying new opportunities.) Shared support services the park offers tenants (education/training, dining facilities, daycare, etc.) will also help build relationships.

"Technology is evolving so rapidly, we should design eco-industrial parks for optimal flexibility, disassembly, and reconstruction. We're moving toward a flexible, modular infrastructure concept. This is a targetable engineering objective. For instance, chemical process and equipment design often enables pulling a few switches and generating different products from the same input stream. This gives resilience to the use of capital equipment in the face of shifting market demand and business cycles.

"This principle of design for flexibility may be the easiest way of communicating the idea of the 'learning system' to engineers and developers. One of the characteristics of a learning system is that you have ease of making and breaking connections as conditions change. This idea can be used both literally and metaphorically in the design of an industrial park."

(Tibbs 1994)

EIP BENEFITS

Benefits to Industry

For the companies involved, the eco-industrial park offers the opportunity to decrease production costs through increased materials and energy efficiency, waste recycling, and elimination of practices that incur regulatory penalties. Increased efficiency may also enable park members to produce more competitive products. In addition, some expenses once incurred solely by individual businesses may be shared by firms in the park. These may include shared waste management, training, purchasing, emergency management teams, environmental information systems, and other support services. Such industrial cost sharing could help park members achieve greater economic efficiency than their stand-alone counterparts.

Benefits to the Environment

In addition to the reduction in many sources of pollution and waste as well as a decreased demand for natural resources, eco-industrial parks will demonstrate in a real-world setting the principles of sustainable development. Most importantly, each park will serve as a working model for future eco-parks and other environmentally sound forms of business operation.

Benefits to Society

The probable enhanced economic performance of participating businesses will make EIPs a powerful economic development tool for communities. Such parks are likely to attract leading-edge corporations and open niches for new local ventures. Both will create new jobs in much cleaner industrial facilities.

Reduction in solid and liquid waste streams will reduce demands on municipal infrastructure and budgets.

EIPs offers government, at all levels, a laboratory for creation of policy and regulations that are more effective for the environment while less burdensome to business.

Eco-Restoration at General Electric

In 1947, General Electric's Medical Systems Division in a Milwaukee suburb discovered notable cost savings from ecosystem restoration. GE brought in a landscape architect to restore an 80 acre tract at this facility. The resulting prairie shows employees and residents something similar to the pre-conquest ecosystem of the region. The project had continuing bottom line benefits, as well. The landscaping cost $300-400 per acre, compared with the $1000/A cost of conventional bluegrass lawns. GE also cut the maintenance costs for mowing, watering, fertilizing and weed control, (which typically run $500/A per year).

Such ecological restoration projects are beginning to play a role in companies with large landholdings, such as Pacific Gas and Electric in California and in public infrastructure projects such as highways.

(Berger 1987)

COSTS, RISKS, AND CHALLENGES OF EIP DEVELOPMENT

Developing an eco-industrial park is a complex undertaking, demanding integration across many fields of design and decision-making. Success depends upon a new level of collaboration among public agencies, design professions, project contractors, and companies locating in the park.

The financial community may be reluctant to support development of an "unproven" approach to industrial parks, however promising it may appear. Some of an EIPs benefits may only become apparent when costs and savings are calculated in a longer time frame than is typical in industrial park financing. On the other hand, if major companies are committed to locating in an EIP, this will start to prove the concept to financiers.

Eco-industrial parks *may* cost more to develop than traditional parks, depending upon the design choices in a project. When this occurs, the additional costs may or may not be offset by savings in operating the park as an EIP, given the payback period acceptable to the developer. Public development authorities may be better prepared to bear this possible increase in development costs than private developers.

Possible innovations in regulation to enable EIP development may not be allowed by regulatory agencies. Even if they are adopted, ideas such as site-wide permitting may be attractive for smaller companies, but could be a disincentive to larger ones. In either case, they will have to be designed so that the whole park will not be liable for possible infractions by one company.

Most companies are not used to working "in community" and may fear the interdependence this creates. On the other hand, many large and small companies see such interdependence as a major source of competitive advantage.

See also the challenges listed above for industrial ecosystems.

ECO-INDUSTRIAL PARK PROJECTS

Dalhousie University's Burnside Project: Industrial Park as Ecosystem

The Burnside Industrial Park in Nova Scotia is the subject of research and development by the School for Resource and Environmental studies at Dalhousie University. An interdisciplinary team at the large site (including 1200 small and medium-sized businesses) has developed principles, guidelines, and strategies for enabling an existing collection of plants to become an industrial ecosystem. The project focuses on several key areas:

- Enabling materials and energy cycling among companies through co-location of businesses, a waste exchange information system, and attraction of scavenger and decomposer businesses that typically use or trade "secondhand" materials
- Relating buildings and the park as a whole to the natural environment, including use of passive solar heating in buildings and use of wetlands to filter runoff and sewage
- Making information available to resident companies on materials used, energy required, and wastes generated in Burnside
- Creating feedback loops within and between companies, and with park management, and regulators.

A software decision-support system, ECOPARK, is under development for use by industrial park managers and business people in parks. It contains data bases on businesses, materials used, technologies, techniques, law and regulation, government assistance, products made from recycled and recovered materials, and case studies. ECOPARK is designed to support businesses in discovering potential markets and sources for by-products and to do cost-benefit analysis of potential trades.

PCSD EIP Demonstration Projects

The President's Council on Sustainable Development (PCSD) adopted the eco-industrial park concept as a basis for demonstration projects and named four communities as sites for these demonstrations: Chattanooga, Brownsville, Baltimore, and Cape Charles.

Chattanooga, Tennessee

Chattanooga has created a sustainable community initiative with broad citizen involvement. It features ecosystem cleanup, an environmental business economic development plan, and identification of four potential sites for eco-industrial parks. These sites include a set of now contaminated properties downtown; a former Army munitions manufacturing facility; a greenfield parcel to be developed for light industry, commercial, and residential; and a decommissioned glass factory in a low-income neighborhood. Funds for planning two of the parks have already been raised.

Brownsville, Texas: An Eco-Park on the Border

A project team from Research Triangle Institute and Indigo Development surveyed selected local companies to identify potential players in an eco-industrial park in this cross-border region. The project's purpose was to build an economic and environmental model to simulate the benefits and costs of an EIP. The process uncovered the possibility of creating a park at the Brownsville Port, with links to other companies in the area. The field research in Brownsville also contributed to Indigo's writing of the fieldbook for development of eco-industrial parks.

The anchors for this *possible* EIP could be an electric power plant and a petroleum refinery (two of the key industries at the by-product exchange network in Kalundborg, Denmark). The power plant would use gas from the refinery and provide it with steam. Other companies would include an asphalt plant (using residual oil from the refinery and steam from the power plant); a wallboard company (using gypsum from the power plant); a tank farm (using steam from the power plant); and a number of others including a water pretreatment plant and an oil recycler.

In the next phase of exploration, the City of Brownsville used an Economic Development Administration technical assistance grant to use computer models to analyze potential energy and materials flows among existing manufacturing plants and to determine regulatory changes needed to enable developing an industrial ecosystem among these firms. The possible EIP at the Port will be considered later.

The Baltimore Empowerment Zone Eco-Industrial Park

The Baltimore Development Corporation (BDC) has led the local EIP initiative by including the concept in the city's successful Empowerment Zone project funded by the Department of Housing and Urban Development. With support from a Cornell University Team, BDC has moved the project forward through a Search Conference to build a vision and broad community support, a planning charrette, and development of a master plan.

The Baltimore project focuses on an established industrial region of the city, the Fairfield district, covering approximately 2000 acres. The area features companies whose production is based on petroleum and organic chemicals. It is a "carbon" economy of oil company marketing sites, asphalt manufacturing and distributing, and divisions of multinational chemical companies making cleaning solutions, herbicides, or plating solutions.

This project goes beyond strictly environmental goals in its new approach to economic development and business organization. It emphasizes a network model of economic and industrial development in which smaller companies collaborate to achieve marketing advantage and develop joint products.

The Port of Cape Charles Sustainable Technologies Park

The Port of Cape Charles Sustainable Technologies Park is located on the Chesapeake Bay at the southern tip of Virginia's Eastern Shore. In the past several decades, Cape Charles and the surrounding agricultural communities within Northampton County have been plagued with a declining population, high unemployment, and an eroding economic base further threatened by environmental degradation of the Chesapeake Bay.

A sustainable county planning process led to creation of an eco-park concept and the acquisition of the land required in the Port area. A community design charrette in 1995 outlined the Cape Charles Principles to guide the development process. The project held a ground-breaking ceremony in October 1996, with a first tenant signed up, a solar building components company that manufactures roof tiles and facades incorporating photo voltaics. The principles discussed in the next section indicate the broad dimensions of the design and development strategy for this project.

The Cape Charles Principles: Prepared by William McDonough + Partners, April 6, 1995, for The Port of Cape Charles Sustainable Technologies Industrial Park

1. The Sustainable Technologies Park will seek to provide support for industrial, job creating opportunities to:
   - ◆ Support existing local enterprises
   - ◆ Attract new ecologically compatible enterprises
   - ◆ Create new ecologically compatible industries
   - ◆ Offer a national model for environmentally sound coastal development
2. The design of the park will encourage the revitalization of the Cape Charles' historic residential, commercial, and industrial landscape. The sense of place embodied by historic Cape Charles, with its culturally and historically rich landscape, architecture, and society provides the keystone for the responsible future development of the town. Working with the historic landscape aims to discourage ruinous sprawl on surrounding open and rural land.
3. The town and the park will serve as a model for preserving and advancing the Eastern Shore's traditional settlement pattern of compact villages and towns surrounded by productive land and water. Consistent with this development pattern, the town of Cape Charles will be reestablished as an integrated employment and residential center for the region.

4. The park will seek to create "family wage" local employment, training, and opportunities for advancement. The design, capitalization, construction, and occupation of the park should seek to achieve local training and employment during every phase. Local companies which assist citizens in saving energy and water should be created immediately to begin the entrepreneurial activity in the community, consistent with the goals of the park.
5. All designs should attempt to be ameliorative or restorative of the coastal watershed's natural systems, to consider and support the fundamental hydrological and biological characteristics of the site's natural state.
6. The park should evidence world leadership in coastal resource management, particularly water quality management. It will consider all scales from the entire watershed to individual water conservation as a model of conservation and remediation. It should discharge only water unavailable for reuse and in a clean and safe condition.
7. As part of a comprehensive resource management program, the park will implement "Industrial Ecologies" and "Zero Emissions" protocols, and establish recycling and composting facilities for the region.
8. Through the information and technologies fostered by the park, the community will seek to become more reliant on natural energy flows. The citizens of Cape Charles have adopted the concept of becoming a "Solar City."
9. The park will serve as a model of integrated agricultural and industrial growth and will foster technologies/enterprises that add value to seafood and agricultural byproducts.

Other EIP Projects in North America

Other communities and regions in North America planning EIPs include: Alameda and Contra Costa Counties; California; Burlington, Vermont; Eugene, Oregon; Londonderry, New Hampshire; Matamoros, Tamaulipas; Minneapolis, Minnesota; Plattsburg, New York; City of Raymond and Skagit County, Washington; Trenton, New Jersey; Tucson, Arizona; and Wake Forest, North Carolina.

These projects range from six to hundreds of acres. Some are basically seeking to link existing companies into by-product exchanges. Others are new industrial park developments with a variety of strategies for improving environmental performance.

International Projects

Internationally, researchers, companies, and/or developers are working on industrial ecosystem projects and some EIPs in the Netherlands, Austria, Spain, Costa Rica, South Africa, and several Asian countries. In the Netherlands, for instance, a team from the Erasmus University Centre for Environmental Studies has been working with a network of established plants in Rotterdam Harbor to develop exchange of energy and material by-products among them (The INdustrial EcoSystem Project, INES). This initiative, started in 1994 under the leadership of Donald Huisingh, also seeks to upgrade environmental performance within participating plants through Cleaner Production technologies. Companies have identified potential exchanges and have conducted a feasibility analysis.

Important initial research conclusions of the INES project are that the processes of building an understanding of the industrial ecosystem concept, and opening channels of communication and trust is very time consuming. The researchers also emphasize the importance of utilizing or creating a supporting entity outside of the participating companies to coordinate the intercompany processes involved (Bass 1997, Huisingh 1997).

See the sidebar on the Zero Emissions Research Initiative for an international program based at the UN University in Japan.

RESOURCES ON INDUSTRIAL ECOSYSTEMS AND ECO-INDUSTRIAL PARKS

Organizations and Electronic Sources

The Baltimore EIP project is covered at Cornell University's Web site: http://www.cfe.cornell.edu/wei/feip.html.

Brownsville EIP project information can be found on the Border Information and Solutions Network. http://www.triplesoft.com/bisn/.

Center for the Environment and School for Industrial and Labor Relations, Cornell University, Ithaca, NY 14853-3901. Telephone: 607-255-8106, Fax: 607-255-4496. Internet: ecr23@cornell.edu or ec42@aol.com. Contact: Ed Cohen Rosenthal. Work and Environment Initiative: http://www.cfe.cornell.edu/wei/. (Extensive discussion of Cornell's industrial ecology team's work on eco-industrial parks.)

Dalhousie University's Burnside Park Industrial Ecosystem Project. Contact: Ray Côté, School for Resource and Environmental Studies, Faculty of Management, Dalhousie University, Halifax, Nova Scotia, B3J 1B9, CANADA. Telephone: 902-494-3632, Fax: 902-494-3728.

DOE Center for Excellence in Sustainable Development: http://www.sustainable.doe.gov/industrial/index.html. (Includes papers on eco-industrial parks and links to demonstration project web sites.)

The Duwamish Coalition industrial ecosystem project: http://www.pan.ci.seattle.wa.us/BUSINESS/DC/default.htm.

Indigo Development, 6423 Oakwood Dr., Oakland, CA 94611. Telephone: 510-339-1090, Fax: 510-339-9362. Internet: elowe@indigodev.com. Web: www.indigodev.com. (Indigo Development is a consulting and research group applying industrial ecology. Its team coauthored this text as well as the *Fieldbook for the Development of Eco-Industrial Parks.*)

President's Council on Sustainable Development: www.whitehouse.gov/PCSD (Includes detailed reports on the PCSD's four EIP demonstration sites: Baltimore, MD; Brownsville, TX; Cape Charles, VA; and Chattanooga TN.)

Zero Emissions Research Initiative, UN University, 53-70 Jingumae 5 Chrome, Shibuya-ku, Tokyo, Japan. Fax: (81-3) 3499-2828, E-mail: foo@hq.unu.edu, Web: http://www.zeri.org.

As part of its Information Project, IIEC publishes *An International Directory of Organizations that Provide Information on Energy-Efficient Technologies.* The directory includes many aspects of energy design beyond efficient technologies. It provides several pages of detailed information on activities and publications of each organization. Many of the groups listed include industrial facility energy design as a central interest.

Bibliographic Resources

Ayres, Robert. 1996. "Creating Industrial Ecosystems: A Viable Management Strategy?" *International Journal of Technology Management*. Vol. 12, Nos. 5/6, Special Issue.

Bass, Leo. 1997. *Cleaner Production and Industrial Ecosystems: A Dutch Experience.* Unpublished manuscript describing the Industrial EcoSystem Project (INES) implemented in the Rotterdam Harbor by a team from Erasmus University working with local industrial managers. E-mail: bass@mil.fsw.eur.nl for publication information.

Berger, John. 1987. *Restoring the Earth*. Anchor Press, Doubleday, NY, pp. 123-4.

Christensen, Jorgen. 1994. "Kalundborg: Industrial Symbiosis in Denmark." In *Proceedings, Industrial Ecology Workshop, Making Business More Competitive*, Ontario Ministry of Environment and Energy, Toronto, Canada.

Côté, Raymond P., et al. 1994. *Designing and Operating Industrial Parks as Ecosystems*. School for Resource and Environmental Studies, Faculty of Management, Dalhousie University. Halifax, Nova Scotia B3J 1B9, Canada. (This volume from Dalhousie University in Nova Scotia, Canada, is an essential resource for EIP teams. On the basis of an extensive literature search and the analysis of Burnside Industrial Park (a large brownfield site), the Canadian team has formulated a set of principles and guidelines governing the design and operation of industrial parks. The handbook's sections include background, waste materials, the development of Burnside Industrial Park, project findings, implementing the ecosystem concept, principles, guidelines, strategies, symbiotic relationships, support systems. and conclusions.

Engberg, Holger. 1995. "The Industrial Symbiosis at Kalundborg." In Gladwin, Thomas, and Freeman, Tara (eds.). *Business, Nature and Society: Towards Sustainable Enterprise*. Richard D. Irwin, Burr Ridge, IL.

Frosch, Robert A., and Gallopoulos, Nicholas E. 1989. "Strategies for Manufacturing." *Scientific American,* Special Edition, September, pp. 144-152.

Gertler, Nicholas. 1994. *Industrial Symbiosis in Kalundborg: Development and Implications*. TB&E Working Paper. Program on Technology, Business, and Environment. Massachusetts Institute of Technology, Cambridge, MA.

Gertler, Nicholas. 1995. *Industrial Ecosystems: Developing Sustainable Industrial Structures*. Dissertation for Master of Science in Technology and Policy and Master of Science in Civil and Environmental Engineering at Massachusetts Institute of Technology, Cambridge, MA. (Gertler's dissertation is the most comprehensive review of conditions for creating intercompany by-product exchanges. He gives a detailed description of the exchange pattern at Kalundborg, the Zero Emissions Research

Initiative, regulatory changes needed, and organizational options for creating and managing such exchanges.)

Kibert, Charles J. (ed.) 1994. *Sustainable Construction: The Proceedings of the First International Conference on Sustainable Construction*. Tampa, FL, November 6-9, 1994) (This 885-page volume contains 90 papers written by authors from 40 countries. The various topics covered include defining sustainability, green initiatives, analytical and assessment tools, economics of sustainability, alternative materials, construction waste studies, and methods in design and construction. It includes many detailed articles on deconstruction and construction materials recycling.)

Knight, Peter. 1993. "Closing the Loop." *Tomorrow, Global Environmental Business Magazine*. Stockholm (article with pictures of Kalundborg, Denmark).

Lowe, Ernest A. 1997. "Creating By-Product Resource Exchanges for Eco-Industrial Parks." *Journal of Cleaner Production.* Vol. 4, No. 4, an industrial ecology special issue, Elsevier Publishers.

Lowe, Ernest, and Evans, Laurence. 1995. "Industrial Ecology and Industrial Ecosystems," *Journal of Cleaner Production*, Vol. 3, No. 1-2.

Lowe, Ernest A., Moran, Stephen R., and Holmes, Douglas B. 1997. *Eco-Industrial Parks: A Guidebook for Local Development Teams.* Indigo Development. (This is essentially the same text as the *Fieldbook for the Development of Eco-Industrial Parks*, drafted by the same authors for US-EPA, with some updating and arranged in a sequence designed for local EIP development teams. Available from Indigo Development See the web site www.indigodev.com for ordering information.) This report includes guidelines on initiating EIP projects and integrating them into broader economic development initiatives in communities; setting environmental performance objectives; planning, financing, and recruitment strategies; and designing management systems for EIPs. It also includes an extensive survey of options for the design of infrastructure, buildings, and support services; redeveloping brownfield sites; detailed cases of EIP and industrial ecosystem projects; and extensive organizational and bibliographic resources.

Martin, Sheila, et al. 1996. *Developing an Eco-Industrial Park: Supporting Research*. Volume 1, Final Report, Research Triangle Institute Project Number 6050, Research Triangle Park, NC. This report includes A Case Study of a Prototype EIP in Brownsville/Matamoros, Regulatory Issues and Approaches for Encouraging Eco-Industrial Park Development, and Technologies Supporting Eco-Industrial Parks. To order, write RTI Information Services, POB 12194, Research Triangle Park, NC 27709.

Nemerow, Nelson L. 1995. *Zero Pollution for Industry, Waste Minimization Through Industrial Complexes*. John Wiley & Sons, NY.

Pauli, Gunter. 1995. "Industrial Clusters of the Twenty-first Century." In Capra, Fritjof, and Pauli, Gunter. *Steering Business Toward Sustainability*. United Nations University Press, Tokyo, Japan.

Pauli, Gunter. 1995. "Zero Emissions: The New Industrial Clusters." *Ecodecision*, Spring, pp. 26-29.

President's Council on Sustainable Development. 1996. *Eco-Efficiency Task Force Report*. (Includes detailed reports on the PCSD's four EIP demonstration sites: Baltimore, MD; Brownsville, TX; Cape Charles, VA; and Chattanooga, TN. Available from PCSD. Telephone: 202-408-5296 or at www.whitehouse.gov/PCSD.)

Schwarz, Erich. 1994. *Unternehmensnetzwerke im Recycling-Bereich*. Fabler Edition Wissenschaft, Wiesbaden, Germany.

Schwarz, Erich J. 1996. "Recycling-Networks: A Building Block Towards a Sustainable Development." International Solid Waste Association, International Congress Proceedings.

Schwarz, Erich, and Steininger, Karl W. 1995. *The Industrial Recycling-Network: Enhancing Regional Development*. Research Memorandum No. 9501. April. Institute of Innovation Management, Karl-Franzens University of Graz, Johann-Fux Gasses 36, A-8010, Graz, Austria. Telephone: 43 316-380-3232, Fax: 38 14 15. Internet: erich.schwarz@kfunigraz.ac.at.

Spurlock, J. M., and Ward, H. C. 1980. *Systems-Integration Requirements for the Synergistic Co-Siting of Industrial Activities*. U.S. Department of Commerce, National Technical Information Service Publication PB81-150294, Washington, DC.

Stopps, Terrance P., and Greenwood, Ellen (eds.). 1994. *Proceedings of the Industrial Ecology Workshop. Making Business More Competitive*. Toronto, Ontario, Canada.

Tibbs, Hardin. 1994. Interview with Hardin Tibbs, a design engineer whose articles on industrial ecology have helped communicate IE's principles and concepts broadly.

Wallner, H.P. 1997. "Regional Embeddedness of Industrial Parks —Strategies for Sustainable Production Systems at the Regional Level." *Journal of Cleaner Production*. Vol. 4, No. 4, an industrial ecology special issue, Elsevier Publishers.

Work and Environment Initiative. 1995. *Fairfield Ecological Industrial Park Baseline Study*. Cornell University, Ithica, NY. (This study was done for the EIP demonstration project in Baltimore.)

8 Strategies for Creating and Implementing IE Are Emerging

INTRODUCTION

While there is growing enthusiasm for industrial ecology, there has been little formal discussion of strategies for developing the field as a whole and putting it into practice. Strategic thinking has tended to focus in sub-areas such as design for environment, regulations, or technology policy. Industrial ecologists discuss broader strategy occasionally in informal settings. The IE literature generally does not reflect this discussion except in side comments. (See IEEE EHSC 1995, Richards and Fullerton 1994, pp. 44-47, and Graedel and Allenby 1995, pp. 293-324.)

In this chapter we will articulate a broad strategy for development of the field, drawing upon informal discussions and the occasional references in the literature. This discussion should be taken only as a proposed strategy to spur further exploration of the subject.

In the Executive Briefing we identified four key thrusts for a self-organizing[1] strategy to develop the field of industrial ecology:

---

[1]See the sidebar to the strategy page in the Executive Briefing for discussion of self-organizing systems.

1. Implementation of many decentralized demonstration projects in diverse industrial and public organizations
2. Application of IE principles in more central institutions, the points of social and economic control, such as the World Bank, regulatory agencies, and standards groups
3. Continuing research on the interactions between industry, the environment, and society; action research on demonstration projects; and development of new IE methods and tools
4. Rapid dissemination of information about IE principles and methods, and about the successes and failures of the demonstration projects.

The systems view of IE requires a strategy that coordinates action within individual firms, across networks of firms, and between industry, the market, the government, and the research community. Several basic principles to support these strategic directions for developing IE are

- ◆ Coordinate a decentralized, self-organizing process of innovation with strong information links across systems and levels of system.
- ◆ Engage the points of social and economic control in the process of developing IE.
- ◆ Design policy through public/private, highest common denominator dialogue, framed by IE understandings and project experiences.
- ◆ Use the leverage of industries at risk from environmental degradation (e.g., insurance, finance, and tourism) to support change in the sectors most responsible.
- ◆ Maintain a strong linkage between people and institutions applying IE and scientists researching IE.
- ◆ Develop industry's leadership role in changing consumer behavior and expectations.
- ◆ Build upon present transforming trends in industry.

EMPHASIZE A DECENTRALIZED PROCESS

Coordinate a decentralized, self-organizing process of innovation with strong information links across systems and levels of system. The World Wide Web has become a powerful resource, supporting the linkage between research centers and action projects.

Some critics fear that the broad, systemic approach of industrial ecology implies centralized, top-down planning and control. Taking this approach would be a disabling error. In ecosystems, control is decentralized and evolution is guided by self-organizing processes. Industrial ecologists emphasize that a similar strategy will be fundamental to evolving their discipline.

The high level of innovation required for realizing IE demands redundancy, decentralization, and self-organization. These are critical qualities in the performance of any complex system.

For example, multiple attempts to create industrial ecosystems and eco-industrial parks are under way in communities in North America and Europe. *See Chapter 7.* These projects have different concepts, strategies, and organizational forms, but all seek to apply basic IE principles. Sharing information and research on their processes and outcomes will support these projects as well as newer ones that emerge. Companies considering participation in an EIP could also learn from each other's experience.

The potentially conflicting self-interest of different industries could add to the pressure for change. Major insurance companies, for instance, would be hit hard by global climate change. A number of them are now exploring strategies for encouraging major clients and firms in which they have invested to reduce greenhouse gas emissions. *See the end of Chapter 5 for a scenario illustrating the role of IE in this area.*

Market forces will drive development of IE approaches in individual companies. For instance, entrepreneurial innovations could challenge existing solutions in energy, transportation, and other industries with high environmental impacts. *See the cases at the end of Chapter 3 and the scenario at the end of Chapter 4.*

Leadership is another important element in this decentralized strategy for developing and disseminating industrial ecology. Early adopters will be creating new standards and new market demand, assuming they gain strategic advantage through demonstrated results.

Strong information links between these decentralized projects will include electronic media, conferences, research reports, and site visits. Progress on the Brownsville

Executives at Turner Construction have launched a sustainable construction initiative based on a systems approach to demonstration projects. When asked what their motivation was, Ian Campbell, director of the initiative, said,

"MARKET! The Turner Executive Team sees that the market includes firms that want green buildings. We need to be able to respond to them. This makes us more competitive because we're offering value adding services in response to a growing niche in the market. We're the largest construction firm in the U.S. and we want to lead the industry, to create the competitive edge in construction."

Texas eco-industrial park, for instance, will be reported on the web site for the Border Information and Solutions Network, a local non-profit.

At a more fundamental level, developing countries could leap-frog industrialized countries in creating industrial ecosystems and the socio-economic patterns needed to support them. The more rapid learning possible in start up enterprises could feed back into the learning process in the industrialized countries. The creation of global sustainable development funds can be seen, not as charity, but as investment in basic R&D that will benefit both worlds.

ENGAGE THE POINTS OF SOCIAL AND ECONOMIC CONTROL

"... the existence of international standards for environmental performance of products, processes, and services, even if advisory in nature, will almost certainly stimulate the adoption of the standards by knowledgeable industrial customers. Once that process begins ... corporations have little choice but to revise their operations to meet these standards lest they lose customers."

(Graedel and Allenby 1995)

Engage the points of social and economic control in the process of developing IE (financial institutions, policy, regulatory, and development agencies, and standards setting organizations, among others).

Decentralized, bottom-up innovation is a necessary but not sufficient condition for an effective IE strategy. There are more "central" points of control where IE's systems approach will be important in setting standards, designing policy and regulations, and evaluating projects for financing.

For instance, World Bank and other development banks could test IE methods for evaluating projects. After initial application in demonstration projects, staff would develop criteria that give precedence to projects designed with systemic analysis of environmental, social, and economic impacts.

The International Standards Organization (ISO), the EcoManagement and Audit Scheme (EMAS), and the International Electrotechnical Commission (IEC) are developing environmental guidelines for industrial processes and products. ISO 9000 quality standards have been widely accepted in industry as a condition for competition in global markets.

ISO is defining standards for sustainable industrial development (ISO 14000). These will include voluntary standards for environmental management systems, life-cycle assessment, products, environmental auditing, and labeling. Since the draft standards indicate a strong emphasis on process, industrial ecology principles and methods could play an important role for companies applying them. IE could be especially important in setting performance objectives and achieving more systemic environmental management systems. Such standards should not block implementation of fundamental large-system changes such as those recommended by Walter Stahel. *See Chapter 4.*

DESIGN POLICY THROUGH PUBLIC/PRIVATE DIALOGUE

Design policy through public/private, highest common denominator dialogue, framed by IE understandings and project experiences.

Industrial ecology helps create a common ground for the different stakeholders in policy development. On the one side, it offers a compelling analysis of the ecological challenges we face. On the other, it acknowledges the need for industry to maintain economic and technical viability.

In Chapter 6 we discuss the movement in regulatory agencies toward working with industry to establish performance measures and a context for self-regulation by industries. The Green Plan section of that Chapter and the description of The Natural Step in Chapter 2 describe two tested processes for engaging all stakeholders in a creative dialogue.

It is important to act with awareness of the implications of the transition to sustainability for the political economy. In the process, all industries will need to map a transition path that acknowledges the present higher environmental burden imposed by some of them.

Continued new findings from research in IE and environmental science and reports from action projects will be valuable inputs to the dialogue. Another party to the dialogue will be economic agencies reforming economic indicators, taxes, and other instruments to reflect Environmental and social costs of industrial activity.

While government responsibilities are shifting, stakeholders need to recognize the unique role and limits of public sector activity in long-term research: keeping the books for the earth and for society, maintaining free flows of information, protecting and restoring habitats, and securing the interests of the whole.

USE LEVERAGE OF INDUSTRIES AT RISK

Use the leverage of industries at risk from environmental degradation to support change in the sectors most responsible.

The immediate self-interest of industries like insurance, finance, and tourism could add to the pressure on other industries for adopting industrial ecology. Major insurance companies, for instance, would be hit hard by global climate change. (Some executives already feel their high weather disaster payouts have been caused by climate change.) A number of them are now exploring strategies for encouraging major clients and firms in which they have invested to reduce greenhouse gas emissions. *See the end of Chapter 5 for a scenario illustrating the role of IE in this area.*

Major U.S. banks have formed the Environmental Bankers Association to work collectively to manage and reduce their environmental liabilities. The EBA initially has been concerned about streamlining the process of brownfield cleanup and redevelopment, since banks have many such properties on their books through mortgage defaults. Association members are now beginning to explore how they can use their resources to support positive initiatives such as eco-industrial parks (Dean Jeffery Tellego 1996, personal communication).

Tourism is another industry with an immediate stake in maintaining ecological health around the planet. In this business the quality of the customers' experience is usually quite dependent upon the environmental quality of the destination. Individual companies and the industry's associations could identify procurement policies, lobbying, and other actions available to promote change in polluting industries. Obviously, tourism companies need to also clean up their own operations, seeking energy and materials efficiency, pollution prevention, and minimal disruption to ecosystems.

LINK ACTION AND RESEARCH

Maintain a strong linkage between people and institutions applying IE and scholars researching IE.

IE research in universities, Federal laboratories, corporate R&D centers, UN research networks, and private research institutes should have strong two-way communication with the decentralized IE projects. This link will also be important to policy, financial, and standard setting institutions.

Action research on IE demonstration projects will advance theoretical understanding while enabling course correction in the projects. Growing scientific understanding of the environmental impacts of human activity will help the leaders in these projects to set relevant objectives.

Consortia of companies—organized by region or by sector—are a likely focus for action research in industrial ecology. Such groups would invest time of staff and money to bring in consultants and academics as needed to develop projects, methods, and tools.

See Chapter 3, The Dialogue Between IE as Research and IE as Context for Action, and Chapter 6, Research and Technology Policy.

DEVELOP INDUSTRY'S LEADERSHIP ROLE IN CHANGING CONSUMER BEHAVIOR

Consumer behavior is clearly responsible for a significant share of environmental damage (Durning 1992). However, patterns of consumption are strongly conditioned by the products companies produce and their marketing strategies. Some business leaders have started exploring the implications of changes in consumer behavior and their own responsibility for leading the market with innovations such as more durable and multifunctional products.

While the subject is still highly controversial, companies like Dow Europe and organizations such as the World Business Council on Sustainable Development have started exploring "sustainable production and consumption" (Victory 1995). A basic theme in these explorations is that eco-efficiency in production must be matched by eco-sufficiency in consumption. The quality movement's focus on value gained by the customer could support firms shifting emphasis from number of products sold to quality of service delivered. An adequate level of efficiency in resource utilization may require fundamental shifts in business mission, not just product takeback and recycling.

See Chapter 4 for discussion of one view of changes in this direction already being tested by companies like Xerox and Agfa Gevaert.

BUILD UPON PRESENT TRANSFORMING TRENDS IN INDUSTRY

We are already in the midst of a major transformation in the industrial system, sparked by technological change, the drive for global competitiveness, and new approaches to management and organization design.

Some Aspects of the Industrial Transformation:

- The ongoing information/communication revolution, including:
  - Automation and computer integrated manufacturing, increasing the capability for customization;
  - Electronic data interchange;
  - Distributed information systems supplanting centralized systems.
- The focus on quality and development of technology and organization to support it
- Dematerialization of product (Buckminster Fuller used the word, "ephemeralization" to refer to this continual decrease in material input to products.)
- Emergence of the practice of concurrent engineering
- Partnering between companies, suppliers and customers, with tight information links and systems and processes integrated across company boundaries
- Successes with high performance, self-managing work teams
- Downsizing of staffs and flattening of organizations.

Each aspect of this transformation presents an opportunity for applying industrial ecology principles and practices.

A fundamental condition for the success of this redesign of the industrial system is developing greater synergy among the many initiatives. The present process of change is often piecemeal, fragmented across both organizational functions and academic disciplines. Corporate change managers fear that separate "functions in silos" in their organizations will never work together effectively. They say the competition between specialties often blocks real change from occurring.

Perhaps the various elements of the industrial transformation itself could be modeled as a cooperative ecosystem. This application of IE could increase synergy among the diverse change efforts while it introduced questions of environmental excellence into each initiative. This would be very much in line with the quest for strategic environmental management in some companies.

RESOURCES ON STRATEGY

During, Alan. 1992. *How Much is Enough? The Consumer Society and The Future of the Earth.* W. W. Norton Co. New York & London.

Flaherty, Margare. World Business Council on Sustainable Development, 160 Route de Florissant, CH-1231 Conches-Geneva, Switzerland. Telephone: +41 22 839 3100, Fax: +41 22 839 3131; E-mail: CompuServe 100277.2732. For information on the working group to address sustainable production and consumption.

Graedel, T.E., and Allenby, B.R.. 1995. *Industrial Ecology.* Prentice Hall, Englewood Cliffs, NJ. Chapters 21-22 include discussion of strategies for developing industrial ecology.

Institute of Electrical and Electronics Engineers; Environment, Health and Safety Committee (IEEE EHSC). 1995. "White Paper on Sustainable Development and Industrial Ecology." (To obtain, contact: IEEE Technical Activities, 445 Hoes Lane, Piscataway, NJ 08855-1331. Telephone: 908-562-3908, Fax: 908-562-1769, E-mail:j.cerone@ieee.org.)

Richards, Deanna, and Fullerton, Ann (eds.). 1994. *Industrial Ecology: U.S.-Japan Perspectives.* National Academy of Engineering, Washington, DC.

SustainAbility Ltd. 1995. *Who Needs It? Market Implications of Sustainable Lifestyles.* Scenarios about the future of sustainable consumption and its implications for business, prepared with support from Dow Europe. The People's Hall, 91-97 Freston Road, London W11 4BD, UK. Telephone: +44 171 243 1277, Fax: +44 171 243 0364.

Victory, Kathleen. 1995. "Focus Report, Why Smart Companies Will Take Part in the Debate on Sustainable Production and Consumption." *Business and the Environment.* Vol. 6, No. 8, August, Arlington, MA.

See Chapter 3, Resources on Organizational Design, for resources such as Pinchot and Pinchot 1993, Senge 1990 and 1994, and Beer 1979.

9 Developing Country Leapfrog

INTRODUCTION

Looking backward from the year 2015

The following future scenario of how a developing country could use industrial ecology to chart its path to sustainability is ***written looking backward from the year 2015****. This scenario is fictional and fairly improbable. The "serious questions" posed at the end indicate just how improbable it is, given the political and economic realities of development institutions and developing countries. However, we use the scenario form to push thinking beyond such "realism" and to conceive pathways to more profound change.*

In the Summer of 2000, Zhat Massz, Economic Development Minister for the semi-tropical country of Qwanin, attended a high level briefing on industrial ecology as a path to sustainable development. He asked the presenter to lead a workshop for himself; the Ministers of Environment, Energy, Transportation, and Agriculture; and their top aides.

Qwanin leadership invited development agencies, corporations, non-governmental organizations, and universities from around the world to work with it as a test tube for sustainable development.

This session led the cabinet to commit to a basic development strategy. Qwanin declared itself a world laboratory for sustainable development. It applied industrial ecology and related systems approaches as the foundation for planning and implementation.

Looking backward from the year 2015

Fifteen years later we see a remarkable fulfillment of this strategy. Qwanin has gone through a rapid process of development, attracting both multinational corporations and innovative entrepreneurs. Local businesses have grown, seizing opportunities offered by industrial ecology. Energy and transportation systems have demonstrated the feasibility and profitability of alternative technologies. The diversified economy is strong and environmental conditions have improved throughout the country. How has this happened?

Initially, Qwanin leadership invited development agencies, corporations, non-governmental organizations, and universities from around the world to work with it as a test tube for sustainable development.

Development banks and major foundations provided funding for a national process to create the initiative's vision and to gather essential economic, environmental, and technical information. The visioning used The Natural Step process (developed in Sweden) to build consensus on objectives from the local to national level. *See Chapter 2 for a description of The Natural Step.*

The first wave of research used basic IE methods to analyze Qwanin's industrial metabolism, tracing the dominant flows of materials and energy through its industries and agriculture. This analysis included identification of the largest inefficiencies in materials and energy use. On this basis, local and U.S. researchers created dynamic input-output models of the existing industrial base and baseline environmental conditions. These models enabled development teams to test different scenarios for development, sector by sector. *See Chapter 3 for discussion of industrial metabolism and input-output modeling.*

Students and faculty throughout the educational system helped fill the many gaps in knowledge of Qwanin's environment. In addition intellectual resources from around the world connected into the project via the Internet and research trips. The country became a magnet for both environmental research and economic development talent, wishing to contribute *and to learn*. Would-be sustainable communities around the world formed sister city relationships with cities in Qwanin, exchanging action teams to further two-way learning.

The national development team created an evolving economic development strategy involving all sectors of the private economy and public infrastructure and services. This strategy enabled each sector to apply industrial ecology in their own development activities. As with the vision building process, this planning involved effective communication between local and national levels and between private and public sector players.

Some critical elements of the overall strategy included:

- Basing development strategies on an evolving understanding of Qwanin's local ecosystems, regional environments, and global natural systems
- Focusing external recruitment on companies with a demonstrated commitment to sustainable development
- Offering development incentives to foreign or domestic companies willing to commit to environmental, economic, and social performance objectives they set within the broad objectives of the national strategy[1]
- Balancing external recruitment with development of new local businesses and support for existing industries
- Supporting local business development in IE related fields.

The last strategic point was implemented through international technical transfer programs and local incubators to support development of firms that help achieve IE goals. These included remanufacturing capital equipment; reprocessing material and chemical wastes to use as feedstocks; repairing, remanufacturing, and disassembling consumer equipment; using energy and water efficient equipment; etc. *See Chapter 5 on business opportunities opened by IE principles*.

Qwanin did not rely solely on development bank funding to implement its plan. Insurance companies, pension funds, foundation program related investment, and social venture investors all played a role. Domestic as well as Asian, European, and North American investors participated.

Energy and transportation infrastructure demonstrates Qwanin's innovative approach to development. These

[1]Massz saw that it would be essential for new industry in Qwanin to compete with standards set by European countries and corporations.

Looking backward from the year 2015

major projects used the industrial metabolism studies and dynamic input-output models extensively to put alternative infrastructure development scenarios through a what-if analysis. System designers then used life-cycle analysis and design for environment tools in planning specific energy and transportation projects.

ENERGY

The country's power grid was poorly developed so the strategy emphasized a decentralized energy system based on a combination of relatively small and clean fossil fuel plants and renewable sources. The diversity and scale of technologies was essential to creating a reliable and resilient system.

The Economic Development and Energy Ministries put out a request for qualifications (RFQ) defining this strategy and inviting energy companies to prove their capability to help implement it. The winning proposal came from an international consortium of small to mid-size companies at the cutting edge of energy planning and technology. This group of companies projected targets for building energy capacity with efficiency higher and emissions of greenhouse gases far lower than Qwanin's performance objectives demanded.

Working with the Energy Ministry, this consortium developed a whole systems plan avoiding investment in large power installations and older technologies.

The strategy emphasized phased development of smaller energy generating modules, as needed.[2] It called for modest improvements to the power grid rather than the massive grid proposed by a large Asian engineering company.

The relatively small fossil fuel plants use advanced technologies to ensure low emissions. They were sited near industrial plants that could use their waste heat. Industrial plants with large waste heat output used co-generation to create electricity for themselves and their neighbors.

The strategy's renewable energy sources included photovoltaics; passive solar design in new buildings and

[2]Such modular systems are more flexible and resilient and require a much smaller initial investment.

Looking backward from the year 2015

homes; small-to-medium hydro projects; wind power; geothermal, and ethanol production and other biomass technologies.

The consortium's plan eliminated the major costs of a traditional solution: large centralized power plants and a heavy power grid. Without these large investments in the project budget the energy system could make cost-effective use of photovoltaic cells (which were not yet competitive with fossil fuels in developed countries).

As an additional benefit, the phased development allowed installation of newer renewable technologies, particularly hydrogen fuel cells, when they became cost-competitive.

Qwanin's energy strategy also called for the government and utilities to issue RFPs for 15 year contracts to acquire photovoltaics for all new government buildings and for new utility installations. With these guaranteed domestic markets, a photovoltaic plant broke ground in 1997 to make PV modules, roof tiles, and film for windows.

TRANSPORTATION

Qwanin also had a relatively weak highway system but a well developed railroad network and several navigable rivers. The Transportation Ministry decided to emphasize these resources and recruit companies able to develop a comprehensive rail and water-based system.

A Japanese electronics company took the lead in assembling a team including local and foreign transport entrepreneurs, a European rail equipment manufacturer, and a U.S.-Japanese advanced automobile joint venture. The electronics company was in the lead because the success of the system depended upon advanced information technology providing seamless service to travelers and shippers.

The electronics firm was also the most advanced in both industrial ecology and general systems thinking. Its executives and designers easily perceived the whole transportation system as an information-driven organism, operating in the industrial ecosystems of the country.

A local entrepreneur joint-ventured with a U.S. company to establish the world's first factory manufacturing hybrid-electric cars and small trucks exploiting the Hypercar concept developed by a team at Rocky Mountain

Institute. The majority of these vehicles operate on a lease or rental basis, as needed for local transportation. Their use is integrated with access to all other means of transportation. Most longer trips and shipments are via rail and boats.

Now, in 2015, Qwanin's transportation system generates lower emissions and higher energy efficiency per mile traveled than that of any other country in the world.

See the end of Chapter 3 for details on the hybrid-electric cars and this integrated transportation system concept.

Leaders in manufacturing and agricultural sectors also used industrial ecology and systems thinking to guide their progress toward sustainability.

MANUFACTURING

The Ministries of Economic Development and Environmental Quality invited Qwanin's existing manufacturing industries to set environmental performance objectives and goals using an IE framework. This voluntary program built upon the vision and consensus established through The Natural Step process.

The ministries also asked companies considering sites in Quanin to go through a parallel goal-setting process. Participation of both domestic and foreign firms was high since a crucial step was analyzing the potential financial and competitive benefits of improved environmental performance.

With some difficulty, national and local economic development agencies implemented the IE concept of industrial ecosystems in established manufacturing regions and new industrial estates. *See Chapter 7 for discussion of industrial ecosystems and eco-industrial parks.*

The stumbling block was achieving a sense of common interest among firms from Qwanin and a mix of other countries. The first pilot project in an older industrial region included local, Korean, Taiwanese, Swedish, and Canadian plants. In spite of a general intention to participate in inter-firm collaboration, it took almost two years to work through the cultural barriers. When they succeeded in this, a systems model enabled them to form a series of two-way energy, water, and materials by-product exchanges.

Looking backward from the year 2015

Developers of new eco-industrial estates learned from the experience of this first attempt and enjoyed the benefits of the growing enthusiasm and competitiveness of Qwanin's community of companies.

AGRICULTURE

A majority of Qwanin's people were still rural, farming their own land or working on corporate plantations. The sustainable development strategy emphasized maintaining a strong local farming economy that would avoid the massive displacement of rural populations to crowded cities experienced in neighboring countries.

Agricultural corporations and small farmers alike were invited to develop environmental and economic objectives for reducing their use of pesticides, chemical fertilizers, and energy intensive cultivation and processing technologies.

Two other important objectives were maintaining biodiversity of the crop genetic pool and the ecosystems in and around farms; and preventing soil erosion. Traditional knowledge of Qwanin's farmers was complemented by that of organic farmers from Europe and the U.S.[3]

Rice plays an important role in local diet and in the country's exports so one specific focus of biodiversity work was forming seed banks of all known varieties and closely related plants. This was a crucial precaution, given this crop's vulnerability to global warming and ultraviolet exposure due to ozone depletion.

Another facet of the agriculture strategy was developing agribusiness support industries focused on integrated pest management products and services; infrastructure for increased use of natural fertilizers (with chemicals used as supplements when needed for yield); light, energy efficient farm equipment; and information technologies that support high yields (such as sensors, geographic information systems, and satellite positioning systems).

Some larger farming operations introduced kenaf and hemp crops as paper mill feedstocks. (Mills had been using rice straw for years.) Other alternative crops included Gyal

[3]Organic farmers can be seen as the original industrial ecologists. For decades they have operated successful businesses as managed ecosystems.

Looking backward from the year 2015

for latex and Jojoba for its high quality oil. Manioc, a food crop that grows well on marginal land, became a feedstock for an ethanol plant.

One other agricultural innovation in Qwanin was creating "edible landscaping" on industrial estates and municipal infrastructure land. Many facilities grew enough food for all their employees in greenhouses, ponds, and orchards. They also used wastewater and heat from industrial processes, demonstrating IE's loop-closing principle. On-site treatment plants and constructed wetlands processed the industrial water.

Early in the project Zhat Massz saw that building Qwanin's telecommunications infrastructure was essential to the process. He wanted the people in every village, town, and city to be able to participate. He knew that bottom-up planning was essential to the success of the broad national strategy.

TELECOMMUNICATIONS

Early in the project Zhat Massz saw that building Qwanin's telecommunications infrastructure was essential to the process. He wanted the people in every village, town, and city to be able to participate. He knew that bottom-up planning was essential to the success of the broad national strategy.

A team of innovative small to mid-size companies in Quanin and the U.S.—partnered with a major European telecommunications company—won the contract. Their winning proposal emphasized initial use of satellite technology to establish the basic communications network across the country; information and communications kiosks for broad public access; and leasing of integrated network systems for company and government offices. The plan called for phased development of wireless and fiber optic channels to replace the initial dependence on expensive satellite communications.

The team of companies saw Qwanin as a major opportunity to test the IE concept of the service economy proposed by Walter Stahel. *See Chapter 4.* Rather than seeking to create a system to maximize sale of information products, the partnership delivered telecommunications services to Qwanin's people, industry, and government. With revenues for service rather than goods as the economic motivator, preferred equipment would be durable, multifunctional, modular, and easily repaired and upgraded. These qualities extend the life of equipment and reduce demand on resources and energy.

POLICY AND REGULATIONS

Looking backward from the year 2015

When Qwanin declared itself a world laboratory of sustainable development, Zenan P'han, the Minister of Environmental Quality, was flooded with calls. It seemed that every environmental administrator in the world wanted to show him how to design sustainable policies and regulations. He politely asked each caller how they applied cybernetics to their regulatory processes.

After the silence or the puzzled questions, P'han would explain that cybernetics is the science of regulation for living and mechanical systems. He invited them to participate in the design of a new regulatory system that applied this science as well as industrial ecology.

In the first two years of design work, P'han moved from stop-gap traditional environmental regulations to a highly effective, information driven system. Industrial ecologists, cyberneticians, and other systems scientists helped his staff develop mechanisms for self-regulation in public and private sector operations. Regular consultations with industry ensured their collaboration. Third-party auditing and monitoring, timely public access to reports, and channels for peer pressure helped keep firms and agencies honest.

This design for self-regulation unfolded within the broader policy development process that set broad national environmental objectives, sector and industry specific objectives and goals, and personal objectives and goals. *See Chapter 6 for general discussion of policy and regulation in an IE framework.*

THE BASIC GUIDANCE SYSTEM

Parallel to these initiatives in various sectors of the economy, the people and leadership of Qwanin entered into a unique dialogue about the *purpose of development*.

From the earliest days they explored three questions never before addressed in a national development initiative:

What is quality of life for our people?

How will this development process improve our quality of life?

How will it improve the quality of our environment, the foundation for our lives?

What is quality of life for our people?

How will this development process improve our quality of life?

How will it improve the quality of our environment, the foundation for our lives?

Looking backward from the year 2015

Qwanin's answers to these questions became the basic guidance system for their development process.

AFTERWORD

In this fictional account of Qwanin it may appear that we are saying industrial ecology is the golden path to sustainability. Our real claim is that it is a strong systems framework for organizing more specific approaches, methods, and tools.

Many of the specific elements of Qwanin's story have actually occurred somewhere.

In the energy infrastructure section, for instance, IE (and basic systems thinking) enabled the design of a new infrastructure that integrated advanced renewable and fossil technologies, efficiency of equipment, and patterns of use. The consortium used dynamic input-output modeling to test competing designs against environmental, technical, and economic criteria.

The project's success came from years of innovation in fields like energy efficiency, renewable technologies, and clean fossil fuel technologies; *and from the combination of these innovations into a whole system tuned to its natural environment.*

Many of the specific elements of Qwanin's story have actually occurred somewhere.

- ◆ Researchers have used industrial metabolism methods to analyze heavy metals dispersion in the Rhine River basin.
- ◆ Economist Faye Duchin has worked with Indonesia to explore alternative scenarios for ending deforestation, using dynamic input-output models.
- ◆ Wind power technologies are now cost-competitive with fossil fuel sources.
- ◆ Photovoltaic technology and passive solar *are integrated into building components and design.*
- ◆ Atlanta has declared itself a solar city and is planning a public photovoltaic acquisition program to increase demand and drive costs down.
- ◆ Companies, communities, and developers in North America and Europe are planning eco-industrial parks and industrial ecosystems.

Perhaps the fictional story of Qwanin could actually become a reality. All it will take will be bringing the pieces together and answering . . .

Looking backward from the year 2015

A FEW SERIOUS QUESTIONS

If you were a leader in a developing country and you dreamed you could realize this vision . . .

- *How would you enlist your country's elite to participate if they believed their interests were more aligned with those of global institutions and markets than the interests of your people?*
- *How would you overcome patterns of corruption that could disable the vision?*
- *How would you handle the burden of national debt your country probably carries?*
- *How would you recruit international lending institutions to your vision and overcome their bias toward traditional, centralized solutions?*
- *How would you identify the corporations really able to help you fulfill your strategies of sustainable development?*
- *How would you compete with the messages of international media defining quality of life for your people?*
- *How would you make each stumble along the way be an occasion for learning?*

10 General Industrial Ecology Resources

Note: Internet addresses change frequently. Additional links can be found at http://www.estd.battelle.org.

ORGANIZATIONS AND ELECTRONIC RESOURCES

Dalhousie University, Burnside Park Industrial Ecosystem Project, Contact: Professor Raymond Côté, School for Resource & Environmental Studies, Faculty of Management, Dalhousie University, Halifax, Nova Scotia, B3J 1B9, Canada. Telephone: 902-494-3632, Fax: 902-494-3728.

IE bibliography page: http://web.mit.edu/ctpid/www/tbe/ iebib.htm (includes a very extensive categorized bibliography).

Indigo Development, 6423 Oakwood Dr., Oakland, CA 94611. Telephone: 510-339-1090, Fax: 510-339-9362. Internet: elowe@indigodv.com. Indigo Development is a consulting and research group applying industrial ecology in communities and business. Its team coauthored this text, as well as the *Fieldbook for the Development of Eco-Industrial Parks*. Web: www.indigodev.com.

Institute of Electrical and Electronics Engineers, Environment, Health and Safety Committee, IEEE Technical Activities, 445 Hoes Lane, Piscataway, NJ 08855-1331.

Telephone: 908-562-3908, Fax: 908-562-1769. Web site for IEEE White Paper: http://www.ieee.org/ehs/ehswp.html.

Journal of Industrial Ecology: http: //www-mitpress.mit.edu/jrnls-catalog/jie.html. Edited at Yale University, the journal advances the goals and objectives set forth in the National Environmental Technology Strategy. Contact Reid Lifset at the Program on Solid Waste Policy, Yale School of Forestry and Environmental Studies, 205 Prospect Street, New Haven, CT 06511-2106. Telephone: 203-432-3253, Fax: 203-432-5912, E-mail: RLifset@Minerva.CIS.Yale.edu.

National Academy of Engineering (NAE), Technology and Environment Program, 2101 Constitution Ave. NW, Washington, DC 20418. Fax: 202-334-1563. Telephone: 202-334-2290. For more information contact Deanna J. Richards, Program Director, at 202-334-1516 or by E-mail at drichard@nas.edu. This program area of the NAE continues to play a central role in the evolution of industrial ecology, focusing on business best practices in environmental management and design, as well as engaging engineers, economists, and ecologists in better defining the relation between economic activity and natural ecosystem health and vitality. To order publications contact: National Academy Press, Telephone: 202-334-3313 or 1-800-642-6242.

National Pollution Prevention Center for Higher Education, University of Michigan, Dana Building, 430 East University Ave., Ann Arbor, MI 48109-1115. Telephone: 313-764-1412, Fax: 313-936-2195, E-mail: nppc@umich.edu. Publishes a number of industrial ecology curriculum guides and a bibliography. University of Michigan web site: http://www.umich.edu/~nppcpub/ind.ecol.html. Industrial ecology resources: http://www.umich.edu/~nppcpub/ResLists/Ind.Ec.html.

Program for the Human Environment, The Rockfeller University, 1230 York Ave., Box 234, New York, NY 10021-6399. Telephone: 212-327-7842, Fax: 212-327-7519, E-mail: phe@rockvax.rockefeller.edu. Web: http://www.rockefeller.edu/phe/index.html. This site includes some papers that can be downloaded and a bibliography of the work of one of industrial ecology's early pioneers, Professor Jesse H. Ausubel, and colleagues.

Robert Frosch, **Project on Public Policy for Industrial Ecology**, the Kennedy School Of Government, Center for Science and International Affairs, Harvard University, 79 John F. Kennedy St., Cambridge, MA 02138. Telephone: 617-495-8132, Fax: 617-495-8963.

The Technology, Business, and Environment Program at Massachusetts Institute of Technology, Rm. E40-236, 1 Amherst St., Cambridge, MA 02139. The program was founded to help companies meet the dual challenges of achieving environmental excellence and business success. Industrial ecology, design for environment, and environmental business organization are among the principal areas of interest. Web: http://web.mit.edu/ctpid/www/tbe/. The site includes an extensive bibliography with some downloadable articles.

US-EPA National Center for Environmental Publications and Information (NCEPI), 26 West Martin Luther King Dr., Cincinnati, OH 45268. Telephone: 513-569-7985. If you know the EPA document number and publication title, you can place an order for single copies. Telephone: 513-891-6561 or at Fax: 513-891-6685.

Zero Emissions Research Institute: http://www.zeri.org.

BIBLIOGRAPHIC RESOURCES

Allenby, Braden R., and Richards, Deanna J. 1994. *The Greening of Industrial Ecosystems*. National Academy Press, Washington, DC. The full text is online at the National Academy Press web site. (You need a browser with frames capability, such as Netscape 3.0 or higher): http://www.nap.edu/readingroom/records/0309049377.html. A collection of articles that build on earlier conceptual work on industrial ecology. The works strive to develop the context of industrial ecology with chapters on energy, wastes, pollution prevention, law, economics, and the role of government, along with examples in which aspects of industrial ecology are already emerging, such as in the auto industry and telecommunications. The book concludes by identifying future roles of universities and institutes; it also details areas where research efforts are needed.

Ausubel, Jesse H., and Sladovich, Hedy E. (eds.). 1989. "Technology and Environment." *Proceedings of the First Conference on Industrial Ecology*. National Academy Press, Washington, DC.

Ayres, Robert U., and Ayres, Leslie. 1996. *Industrial Ecology: Towards Closing the Materials Cycle.* Edward Elgar Publishers, London, UK.

Côté, R. P. et al. 1994. *Designing and Operating Industrial Parks as Ecosystems*. School for Resource and Environmental Studies, Faculty of Management, Dalhousie University, Halifax, Nova Scotia B3J 1B9, Canada.

Erkman, Suren. 1995. *Ecologie industrielle, metabolisme industriel, et societe d'utilisation*. A major study supported by the Foundation for the Progress of Humanity, Paris. Dr. Erkman has gathered information on the evolution and application of industrial ecology in Europe, North America, and Asia. An English translation is scheduled to be available by fall 1995.

Erkman, Suren. "Industrial Ecology: A Historical View." *Journal of Cleaner Production*. Vol. 4, No. 4 (an industrial ecology special issue). Elsevier Publishers.

Garner, Andy, and Keoleian, Gregory. 1995. "Industrial Ecology." National Pollution Prevention Center for Higher Education, Ann Arbor, MI (see organizations above for contact information to order).

Graedel, T. E., and Allenby, B. R. 1995. *Industrial Ecology*. Englewood Cliffs, NJ: Prentice Hall. This is the first university textbook on industrial ecology. It offers a strong overview of the environmental concerns generated by industry's impact on natural systems, ranging from global to local scales. In terms of methodology, the work focuses on life-cycle assessment and design for environment, giving a very detailed description of the matrix approaches developed by these authors. It does not cover the work of Ayres, Duchin, or Stahel. The text concludes with forward-looking topics, including organizational opportunities and constraints, standards, and new enabling technologies.

Hall, Charles S., Cleveland, Cutler J., and Kaufmann, Robert. 1986. *Energy and Resource Quality: The Ecology of the Economic Process*. John Wiley & Sons, New York. The expression "industrial ecosystems" and, more importantly, the full concept of industrial ecology, appears in the Preface on page xiv. It is probably one of the very first occurrences of this expression in the published literature.

Institute of Electrical and Electronics Engineers, Environment, Health, and Safety Committee. 1995. "White Paper on Sustainable Development and Industrial Ecology." (See IEEE organizations for access.) World Wide Web site for IEEE White Paper: http://www.ieee.org/ehs/ehswp.html.

JETRO, "Ecofactory—Concept and R&D Themes." *New Technology, FY 1992*, Japan External Trade Organization, Tokyo, Japan. Special issue based on work of the Ecofactory Research Group of the Mechanical Engineering Laboratory, Agency of Industrial Science and Technology.

Journal of Cleaner Production. 1997. Special Issue on Industrial Ecology. Vol. 4, No. 4. Elsevier Publishers.

Keoleian, Gregory A. (ed.). 1997. *Industrial Ecology of the Automobile: A Life Cycle Perspective*. Society of Automotive Engineers.

Meadows, Donella, Meadows, Dennis, and Randers, Jorgen. 1992. *Beyond the Limits: Confronting Global Collapse, Envisioning a Sustainable Future*. White River Junction, VT, Chelsea Green Publishers.

National Academy of Sciences. 1992. *Proceedings.* February 1. An NAS colloquium on industrial ecology in May 1991, including papers by industry, government, and academic contributors. 2101 Constitution Avenue NW, Washington, DC 20418. Sales: 202-334-2525.

O'Rourke, Dara, Connelly, Lloyd, and Koshland, Catherine. 1996. "Industrial Ecology: A Critical Review." *International Journal of Environment and Pollution*. February.

Papanek, Victor. 1973. *Design for the Real World, Human Ecology and Social Change*. Bantam Books, NY. Chapter 9, The Tree of Knowledge: Bionics, calls for the field that is now emerging as IE and DFE.

Richards, Deanna J. (ed.). 1996. *The Industrial Green Game: Implications for Environmental Design and Management*. National Academy Press, Washington, DC.

Richards, Deanna, and Fullerton, Ann (eds.). 1994. *Industrial Ecology: U.S.-Japan Perspectives*. National Academy of Engineering, Washington, DC.

Socolow, R., Andrews, C., Berkhout, F., and Thomas, V. 1994. *Industrial Ecology and Global Change*. New York: Cambridge University Press. Focuses on how humankind can continue to industrialize without disrupting and destroying natural ecological systems. Directed toward readers who already have an understanding of the importance of this issue and consequently have the desire to participate in effectively implementing appropriate strategies. Five main sections discuss (1) the industrialization of society, (2) the main natural systems cycles, (3) toxic chemicals in the environment, (4) industrial ecology in firms, and (5) policy making in the context of industrial ecology. The articles address critical issues such as recycling, solar energy, chemicals in agriculture, industrial innovation, and international perspectives.

Watz, Jill, et al. (ed.). 1996. *International Journal of Environmentally Conscious Design & Manufacturing*. Vol. 5, Nos. 3-4. ECM Press, Albuquerque, NM. Special issue on environmentally conscious design and manufacturing at Lawrence Livermore Laboratory.

Index

T

U, V, W, X, Y, Z